Red Sea
Invertebrates

To Paula

**The following contributors to this volume
are gratefully acknowledged:**

Major photographic sources

F. Jack Jackson
J. David George
Horst Moosleitner
Jim Doran
Frank Nobbe
Peter Vine

Additional photographs

Ashod Francis
John Murray
S. Hall
John E. Randall
Tony Fleetwood-Wilson

Design and Illustration

Jane Stark

Red Sea Invertebrates

Dr Peter Vine

Illustrated by Jane Stark

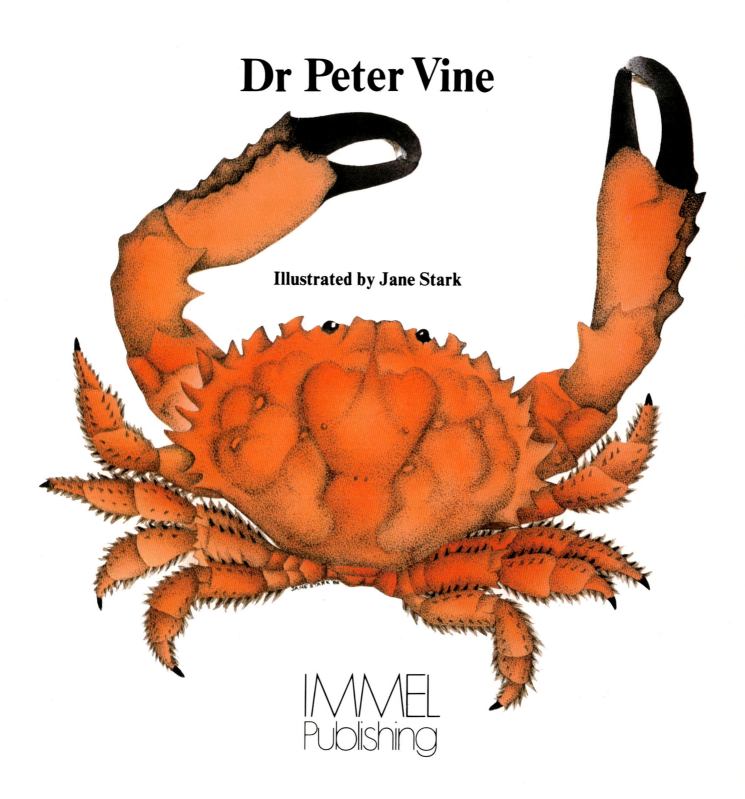

IMMEL
Publishing

Phototypeset in Times Roman by
Lithoset Ltd., Dublin, Ireland.

Printed and bound in Japan by
Dai Nippon Printing Co., Tokyo.

Design and Illustration by
Jane Stark, Connemara Graphics, Ireland.

First published in Great Britain in 1986 by
IMMEL Publishing
Ely House, 37 Dover Street,
London WIX 3RB.

Cataloguing in Publication Data

Vine, Peter
Red Sea Invertebrates
1. Marine Invertebrates — Red Sea
I. Title
592.092 : 733 QL137

ISBN 0907151 - 11 - 6

F. JACK JACKSON

CONTENTS

1. INTRODUCTION

Marine biology as a scientific discipline covers a multitude of skills and can be subdivided into a great number of specialist subjects. Whichever field one follows however, there is a basic need for accurate taxonomy of the marine life one studies. Writing this book has presented the author with an opportunity to review the invaluable contributions to our knowledge of Red Sea marine-life made by 18th and 19th century scientists. They were responsible for discovering, describing, naming and classifying a great many of the species mentioned and illustrated in this text. The first intensive studies of coral-reefs took place in the Red Sea because of its accessibility to European scientists, and for this reason certain groups of animals which have been poorly studied in other coral-reef regions are relatively well known in the Red Sea. Prior to this however, there has not been a single book yet published which deals in any detail with the identification of the main range of Red Sea invertebrates. Until quite recently, the same claim could have been made with regard to Red Sea fishes but that gap has been excellently filled by Immel's publication of "Red Sea Reef Fishes" with photographs and text by Dr J.E. Randall.

Basket star: *Astroba nuda.*

There are a number of problems associated with compiling a comprehensive account of Red Sea invertebrates. Most early scientists worked with dead and preserved creatures. Today's marine biologists have the task of matching their descriptions with live animals. In some cases the differences in form and colour between the two conditions are so enormous that positive identification enters a process of elimination and comes dangerously close to a hit or miss level of accuracy. The advent of SCUBA diving and, with it, underwater photography have brought a new dimension to marine science and in particular to studies on taxonomy, ecology and ethology of marine-life. The Red Sea has been a focus of attention for many young and enthusiastic marine scientists and amateur naturalists who have been entranced by its clear warm waters and abundance of marine life. In consequence, a large number of invertebrates have been photographed in-situ underwater and we are faced with the task of matching these perfectly displayed live animals with a scientific name derived from descriptions of preserved animals. An additional problem one faces in preparing an account such as this one is that of sheer numbers of species. There are well over a thousand species of Red Sea molluscs; around two hundred recorded corals and other groups are equally well represented. It is beyond the scope of any single publication to describe all the recorded species.

Despite its vast range of species however, the Red Sea does offer an opportunity to make intensive studies on certain groups in a well delineated geographic area and one in which the coral reefs are more easily reached than in many other regions. In this book I have used the term "Red Sea" to include the main basin, commencing in the south at the Straits of Bab al Mandeb and including its northerly extremities of the deep Gulf of Aqaba and shallow Gulf of Suez. In terms of distribution of species, this forms a single area in which sea temperature seasonal ranges related to latitudinal differences probably exert a more significant influence on the geographic range of represented species than do any physical barriers between the main basin and the two gulfs. It must be added however that in addition to being cooler, the shallow Gulf of Suez presents more sandy marine habitats and its waters are more turbid than those of the deeper regions. This also affects the distribution of many invertebrates.

Red Sea marine-life is essentially a modification of the Indo-Pacific species assemblage. In most groups, the majority of species also occur in the Indian Ocean and there are relatively few species whose distribution is confined only to the Red Sea and are thus regarded as Red Sea endemics. It is however a measure of the degree of isolation of the Red Sea fauna from that of the Indo-Pacific that endemic species do exist, and that they do so in higher numbers than one would find in more open regions such as the western Indian Ocean.

The reader will be well aware that taxonomy of marine life is based upon the Linnaean system of classification which utilises Latin names. Some popular guide books have attempted to steer away from use of these scientific names in favour of common names. In most cases this causes further confusion since the same species can have half a dozen different common names or, worse still, the same popular name can be applied to several different species. There is however a place for such easy to pronounce names providing the scientific labels are not abandoned. This is doubly true for a region such as the Red Sea where Arabic, English and Italian are all spoken together with several local dialects.

In addition to adopting the correct scientific nomenclature of species, the book classifies them under their relevant taxa and where possible major characteristics of these are mentioned and a list of Red Sea species from each family is given. It is hoped that these check-lists, together with reference to scientific publications, will aid in identification of those species which are not separately illustrated in this volume.

The preparation of this book has been considerably aided by reference to numerous publications and while these are separately listed in the reference section and their authors are duly acknowledged, I should like to make particular reference here to David and Jennifer George's book: "Marine Life" (published by Harrap) and to Doreen Sharabati's recent colour guide entitled "Red Sea Shells" (published by KPI) both of which I thoroughly recommend. I should also like to pay special tribute to illustrator and designer Jane Stark who has greatly added to the value of the text and photographs by her meticulous art-work and by her very special flair for lay-out.

The Red Sea offers the amateur diver and professional scientist some of the World's best opportunities for study of coral-reef marine-life. This book is a testimony to the incredible abundance of species which occur in its waters. It should also serve as a reminder that this unique biotope deserves our greatest care in ensuring its preservation. Red Sea reefs are among the World's greatest natural wonders. Long may they remain so.

Opposite: Hormothid anemone.
Below: urochordate, *Didemnum moseleyi*.

J. DAVID GEORGE

2. SPONGES

PHYLUM: PORIFERA

Coral-reef sponges present a number of identification problems for non-specialists. Red Sea sponges have been relatively well surveyed but most references date back to the early part of this century or to the late 19th century and descriptions are based upon dried specimens and upon a microscopical examination of their structures. These earlier studies concentrated almost entirely on sponge taxonomy and while they recognised that environmental conditions could lead to considerable variation in sponge form, they are of relatively little assistance in providing field identifications of sponges to underwater divers. The basic sponge structure comprises a collection of cells which enclose a system of canals and chambers which open to the exterior by small pores. The sponge is supported by a skeleton which may be formed by calcareous or siliceous spicules and by a matrix of spongy fibres (spongin). The taxonomy of different sponge species is based upon their skeletal arrangement rather than on their external form. The following account deals only with some selected shallow-water species*.

Above: Yellow sponge, *Ircinia* cf. *ramosa* with brown sponge, *Theonella conica*.
Opposite: Calcareous sponge, *Clathrina* sp.

Class: Calcarea: Calcareous Sponges

Skeleton formed by calcareous spicules. They are relatively small sponges, generally less than 10cms in height, with vase-shaped, cushion-shaped or branching growth forms. Spicules are usually separate from each other and have three to four rays radiating from a central point. Simple rods may also occur.

Subclass: Calcinea

Order: Clathrinida

FAMILY: CLATHRINIDAE
Clathrina coriacea. A cosmopolitan, small calcareous sponge which may form a lattice-work of anastomosing tubes, particularly where strong water movement occurs. In still water its colonies are more upright (up to 5cms high). It prefers shaded habitats.

Clathrina tenuipilosa forms upright, branching, tubular colonies which arise from an anastomosing network of tubes spread over the substrate which is usually the shaded underside of coral boulders in shallow water or on shaded crevices formed by microatoll formations in the shallow lagoon. Other species include *C. ceylonensis; C. darwinii* and *C. poterium.*

Subclass: Calcaronea

Order: Sycettida

FAMILY: HETEROPIIDAE
Grantessa hastifera was originally described by Row (1909) from a single specimen collected at Suez. It is recorded by Nobbe (personal communication) from Aqaba. It has a small syconoid form with a single apical osculum.

Grantessa zanzibaris is also recorded by Nobbe from Aqaba.

*The author is greatly indebted to Frank Nobbe for assistance in providing up-to-date classification of Red Sea sponges and for his excellent sponge photographs.

Black sponge

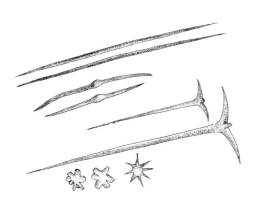

Figure 1: Spicules of *Geodia micropunctata*

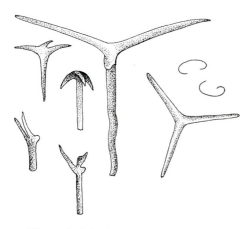

Figure 2: Spicules of *Paratetilla bacca*

Fragment of *Lobophyllia* coral smothered by orange sponge.

Neptune's Goblet Sponge

Class: Demospongiae: Non-calcareous Sponges

Skeleton formed by siliceous spicules, spongin fibres or a combination of both. In a few species there is no skeleton. Spicules usually four-rayed (never six-rayed). Body variable in size and shape and usually without a central large chamber. The internal canal system possesses small chambers which are lined by flagellated collar cells and a few larger canals provide a passage for water to pass out via the osculum.

Subclass: Homoscleromorpha

Skeleton may be absent, or consists of small, similar sized spicules which are usually three-rayed (also two or four).

Placortis simplex is a yellowish white sponge which occurs on mud flats at, for example, Suez.

Subclass: Tetractinomorpha

Relatively large sponges, up to 50cms or more in diameter; skeleton comprises network of four-rayed megascleres and/or one-rayed megascleres. Star-shaped microscleres are often present. Usually without spongin fibres.

Order: Astrophorida

Porifera with skeletons formed by a combination of spicules normally comprising star-shaped microscleres, together with a partial framework of radiating megascleres which converge in the sponge centre.

FAMILY: GEODIIDAE
Geodia micropunctata is irregular and massive in shape; approx 10cms long and 3cms thick. Surface white and perfectly smooth (i.e. without any projecting spicules); pierced by a number of small circular oscula. Spicules are illustrated in figure 1. This species occurs on floating buoys and on suspended pearl oysters and similar substrates.

FAMILY: EPIPOLASIDAE
The family has monaxon megascleres and star-shaped microscleres. It is represented by *Jaspis johnstoni* which is black at the surface and bright yellow internally; and by *Jaspis reptans* which has a smooth dark brown surface (lacking visible orifices) and is light brownish yellow internally. Other species include: *Asteropus simplex* and *Epipolasis aqabaensis*.

Order: Spirophorida

This order consists of globular sponges in which the skeleton is formed by a radiating network of megascleres and spiral microscleres.

FAMILY: TETILLIDAE
Paratetilla bacca is a globular sponge, about 2.5cms in diameter, brownish with a radiating skeletal arrangement. Surface of sponge is slightly rough as a result of a few projecting spicules. Oscula irregularly scattered over the surface (about twelve) and 1-2mm in diameter. Spicules as illustrated in figure 2.

Order: Hadromerida

Sponges in which the skeleton is formed by a radiating system of single rayed megascleres (frequently knobbed at one end), and sometimes including microscleres which are star-shaped or similar.

FAMILY: TETHYIDAE
Members of this family have a strongly developed fibrous cortex and radially arranged skeleton.

Tethya seychellensis is golf-ball like in form, 2-3cms in diameter with a single osculum situated at the summit of a very small papilla. Quite common in shallow sheltered areas where it attaches to hard substrates.

FAMILY: SUBERITIDAE

Suberites clavatus is sub-globular, reddish-yellow or orange; approx. 1.5cms in diameter; centre of sponge is occupied by a large cavity divided into sub-sections by spicular walls which are approximately 1mm thick. The cavity itself is about 7-8mm in diameter. The surrounding smooth envelope is dense and fibrous. There are no visible orifices. Spicules are illustrated in figure 3. This is a relatively common shallow water species which encrusts many surfaces and is often found among littoral mangroves.

Terpios viridis is an important member of the coral-reef sponge fauna in the Red Sea and was originally described by Keller (1891) from specimens collected on corals in the *Stylophora* zone on reefs close to Suakin. During ecological surveys of Red Sea reefs in 1975 the author noted large tracts of a thin grey to blue-grey, slimy sponge, literally blanketing sections of reef. Closer inspection revealed that the culprit (identified as *Terpios viridis*) was successfully growing over live corals and thus killing them. A series of observations on particular corals and time-lapse photographs of the sponge coral interface clearly demonstrated how rapidly this sponge can spread and kill living corals. At one site where regular dives were made, the sponge extended its coverage to an area of 400 square metres during the summer months. This was at a depth of 15 to 20 metres and along a 40 metre length of reef on which all smothered corals were killed. The phenomenon was being simultaneously investigated at Guam in the Pacific Ocean (Bryan, 1973) where *Terpios* sp. was shown to grow at a rate of 2.3cms per month over live colonies of *Porites lutea*. It was clearly demonstrated that the *Terpios* has a toxic effect on live corals, causing retraction of their tentacles and closure of polyps "mouths". In the Red Sea growth rates of sponge colonies encrusting reef-corals seem to vary on a seasonal basis with maximum growth occurring from May to October. A few corals such as *Galaxea* appeared to be more resistant to the sponge than others but no species was completely immune from the smothering effects of rapid sponge growth. Monthly photographs revealed that a much slower growth-rate occurs in winter when the sponge-coral interface remains more or less static. The study indicated at least fifteen sponge species which were active in growing over live corals and the field of sponge-coral ecology is one where continued research is likely to produce many interesting insights into our understanding of reef development.

In his study of Red Sea sponges, collected mainly from the Gulf of Aqaba, Frank Nobbe records the following species from this family: *Laxosuberites cruciatus; Pseudosuberites andrewsi; Suberites carnosus; S. kelleri* and *Terpios granulosa.*

FAMILY: SPIRASTRELLIDAE

Includes the genus *Spirastrella* with the type species *Spirastrella cuncatrix* which possesses a peculiar type of spicule (called spinispirae). Megascleres are tylostyles, more or less radially arranged. Some species bore into coral-rock and thus cause its disintegration.

Spirastrella cuspidifera is a yellowish grey (changing to yellow after death) sponge which spreads over shallow-water coral rocks, giving the appearance of a thin covering but does in fact extend into the rock since its cavities are arranged vertically. It may sometimes give the appearance of a massive growth form after the substrate has disintegrated. A spirally arranged pattern of water collecting channels lead to the oscular and hence its characteristic surface appearance.

Spirastrella inconstans has a variable growth form described as globose, meandrine and digitate. The first two are usually associated with sponges found growing on coral rock, while the latter is found on sponges of this species which sometimes carpet sandy bottoms. The yellow sponge fills the interior of the coral rock, disintegrating it to the maximum and papillae jut out through the surface and spread gradually in the form of a relatively thick disc. Once the rock has disintegrated the papillae may assume a massive form. Incurrent papillae are usually located where there is some protection from silt. Some specimens, after

Tubular sponge, *Acervochalina* sp.

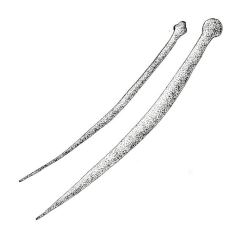

Figure 3: Spicules of *Suberites clavatus*

Mycale fistulifera

J. DAVID GEORGE

Spirastrella coccinea

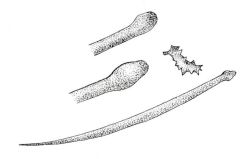

Figure 4: Spicules of *Spirastrella inconstans*

Red sponge, *Cliona vastifica*

J. DAVID GEORGE

disintegrating the substratum may become free-living sponges, not directly dependent upon their boring activity. Spicules are illustrated in figure 4. This species occurs in a wide range of sheltered shallow water habitats, including among mangroves.

FAMILY: CLIONIDAE

A widely distributed family of boring sponges including about fifteen genera. There are approximately 165 species in the genus *Cliona* only.

Cliona celata was first described from a specimen living on an oyster shell. In the Red Sea it also occurs on bivalves (especially pearl oysters) and it bores into their shells in a similar fashion to that observed on European oysters. An early comment on its effect on pearl shells was made by Cyril Crossland who worked at Dunganab Bay Pearl Oyster Farm in the early 1900's. He wrote as follows:—

"The Clione forms a sheet on both surfaces of the shell, a state of things I do not remember seeing before. In this connection note that in many old pearl shells the Clione destroys the outer part of the shell completely, but is unable to penetrate the inner part of the shell and dies off. Many shells show that they were attacked badly some time previously and have recovered."

It is also responsible for disintegration of coral rock into which it bores. Living specimens vary from green to yellow in colour. Incurrent and excurrent papillae are flush with, or slightly protruding from, the surface. Galleries inside the substratum are 2-5mm in diameter and communicate with each other via openings of 1-2mm diameter. Extension of the sponge colony within the substrate is by stolons, which spread irregularly on the peripheral part of the coral rock, and the sponge may penetrate up to 30cms into the rock.

Cliona orientalis spreads on the surface of the coral and never grows beyond it to create a massive form. It is always deep brown at the surface with a pale yellow interior. Contractile oscules (2-5mm diameter) are scattered irregularly on the surface, which is somewhat rough to feel as a result of irregular bands of spicules which project in a brush-like pattern. It may cover an area up to 2 metres or more in diameter, spreading over coral rocks. It can be readily identified as a result of its colour.

Cliona viridis. This bright yellow *Cliona* species bores into coral rock. Papillae are small (1-3mm diameter and height). Galleries inside coral are 3 to 10mm. Papillae are hard and leathery.

Cliona vastifica is recorded as the dominant shallow water *Cliona* species near Aqaba.

Order: Axinellida

Erect sponges, frequently with a branching growth form up to 50cms or more in height. Colour is generally yellowish-orange. Skeleton consists of a longitudinal axis of one-rayed megascleres from which a radiating skeleton arises or, alternatively, with axial framework composed of spongin fibres.

Acanthella aurantiaca generally grows in the form of irregular adjacent lamellae (about 4mm thick), about 8-10cms in height. It is particularly abundant in shallow, sheltered areas such as in Suakin harbour. Spicules are illustrated in figure 5. *Acanthella carteri* is also recorded and illustrated on page 20.

Phakellia palmata forms a frond-like lamella which grows upright on a narrow "stalk", which is about 1cm high and 6mm in diameter. Main body is a broad flat lamella (approx. 5cms in diameter) from the edge of which a number of very short conical processes arise. At the summit of each of these are the oscula which are thus arranged around the edge of the sponge lamella. Surface is coarse and uneven. Colour is dark brown; texture is firm and tough. The only spicules present are styli (see figure 6) and these are extremely abundant.

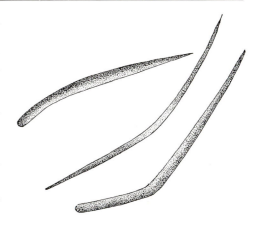

Figure 5: Spicules of *Acanthella aurantiaca*

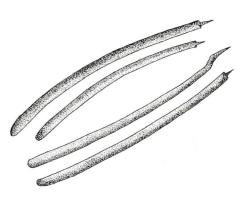

Figure 6: Spicules of *Phakellia palmata*

Cliona vastifica

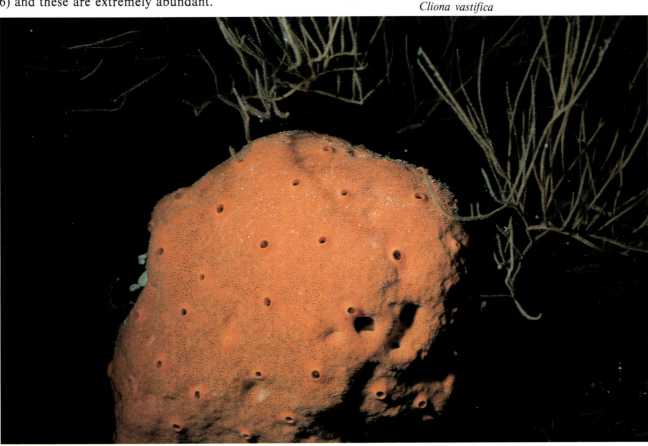

J. DAVID GEORGE

Figure 7: Spicules of *Hymeniacidon calcifera*

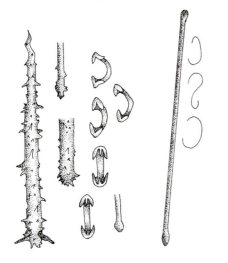

Figure 8: Spicules of *Myxilla cratera*

Blue sponge: *Hymedesmia* sp.

Subclass: Ceractinomorpha

Sponges in which the skeleton is formed by spicules and spongin fibres or by just spongin fibres. Megascleres always have one ray (never four) and most microscleres have the shape of curved hooks, never star-shaped.

Order: Halichondrida

These are encrusting sponges in which the skeleton contains a relatively small amount of spongin. Spicules are normally megascleres which are scattered throughout the body.

FAMILY: HYMENIACIDONIDAE

Hymeniacidon calcifera forms thin sheets of sponge (about 5mm thick) over coral and shell rubble. Surface is irregular and covered with slight prominences and depressions. Oscula are numerous, small and scattered over the surface. Spicules are needle-like (figure 7).

Order: Poecilosclerida

This is a large group of sponges which have skeletons in which megascleres may be joined together by spongin fibres. Microscleres are numerous and hooked. Families include: *Mycalidae; Latrunculiidae; Biemnidae; Crellidae; Myxillidae; Hymedesmiidae* and *Clathriidae.*

FAMILY: MYXILLIDAE

Myxilla cratera is a cushion like mass, rather irregular in shape, of varying size, attached to substrate by wide base. Surface is covered with small crater-like projections which are pore areas. Oscula are small (1.5 to 2mm in diameter). Colour yellow; texture firm. Spicules of varying kinds are illustrated in figure 8.

Yellow sponge: *Leucetta chagosensis*

Pink Tube Sponge: *Siphonochalina siphonella*

Siphonochalina sp.

Tedania anhelans. A large sponge with subspherical, cushion-like form, diameter approx. 12cms with height 8cms. Upper surface foliaceous. In life the colour is vermillion. Spicules are illustrated in figure 9. Note the minutely spined swollen ends of the tylota spicules. This species often grows on live corals and on dead basal portions of live colonies as well as among mangroves.

FAMILY: CLATHRIIDAE

Microciona atrasanguinea forms thin orange plates which are covered in low cone-shaped elevations. Spicules are shown in figure 10.

Order: Haplosclerida

The sponges in this order possess a well developed, reticulated skeleton formed by spongin fibres.

FAMILY: HALICLONIDAE

Adocia dendyi is a bright yellow sub-massive sponge with large oscula. Abundant on shaded surfaces in shallow water.

Cribochalina sp. is a long cylindrical vase-shaped sponge with irregular outer surface formed by numerous finger-like projections.

Siphonochalina siphonella: Pink Tube Sponge.
Orange or pink colonies in the form of parallel branching tubes. No Spicules present and the skeleton is formed from spongin. Unlike many sponges this species is relatively light-loving and is found exposed in quite shallow water near the reef crest.

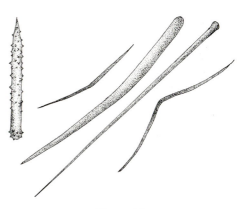

Figure 9: Spicules of *Tedania anhelans*

Figure 10:
Spicules of *Microciona atrasanguinea*

Dysidea cinerea growing on gorgonian sea-fan. Opposite, top: Red sponge: *Latrunculia corticata;* bottom: sponge: *Grayella cyathophora.*

Order: Dictyoceratida

Large, irregular sponges whose skeleton is formed of laminated spongin fibres. No siliceous spicules. Includes the commercial bath sponge.

FAMILY: SPONGIIDAE

Spongia officinalis are large, rounded, sponges with black surface and yellow inside. Usually around 10 to 15cms in diameter. Several cone-like lobes arise from the surface and each has an osculum at its summit. Overall surface of the sponge is uneven. Found in moderate depths, especially where there is a long-reef current.

FAMILY DYSIDEIDAE

Dysidea herbacea can be extremely common and forms "pastures" in some shallow sandy areas such as around the Dhalak archipelago. It has a spongy texture but an irregular spiky surface. *Dysidea cinerea* is also present.

Orange boring sponge attacking coral and causing its death.

FAMILY: VERONGIIDAE

Verongia mollis forms upright, roughly cylindrical processes which are joined together basally. The surface, which is irregularly sculptured with protuberances and depressions, also has a regular distribution of cones which are 1-2mm high and 6-8mm apart. These mark the point where skeletal fibres reach the surface. Oscula are irregularly scattered. Colour: a striking dark green on the outside and internally yellow. Skeleton formed by a few pitted fibres. It occurs on open sandy or muddy bottoms in sheltered areas, at about 8m depth.

Order: Dendroceratida

These have arborescent skeletons formed by non-laminated spongin fibres, or else they lack a skeleton altogether. Siliceous spicules or extraneous material are both absent from the supporting skeletal structure.

FAMILY: APLYSILLIDAE

Aplysilla lacunosa is a small sponge, violet to red in colour, which is common in shallow water, often in association with *Stylophora* corals.

Mycale dendyi

Chondrillastra mixta

Acanthella carteri
Hymedesmia lancifera

Cliona viridis

SPONGE PHOTOGRAPHS AND IDENTIFICATION BY FRANK NOBBE

Theonella swinhoei

Leucetta chagosensis

Clathrina coriacea
Psammaplysilla purpurea

Mycale fistulifera

SPONGE PHOTOGRAPHS AND IDENTIFICATION BY FRANK NOBBE

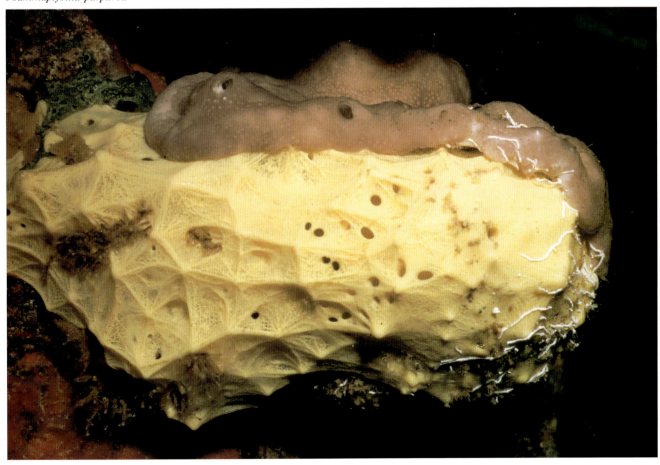

3. COELENTERATES I.

PHYLUM COELENTERATA

Class: Hydrozoa

Red Sea Hydrozoans include small feathery hydroids, stony coralline forms and a number of species which lack polyp stages and exist as free-swimming medusae.

FAMILY: MILLEPORIDAE: Fire-Corals

Fire-corals are a consistent feature of the reef-crest and shallow reef-face on Red Sea coral-reefs. Their relatively smooth, hard, bright yellow growth forms are usually sufficiently conspicuous for divers to avoid brushing their stinging surfaces. Four species are recorded from the Red Sea and Gulf of Aqaba.

Fire coral: *Millepora dichotoma*.
Opposite, insets: Fire coral: *Millepora dichotoma;* main picture: *Millepora dichotoma* and *M. platyphylla.*

Millepora platyphylla has a plate-like form whose surface is covered with small conical "warts". Each wart measures about 2mm and at the centre of its top is a large pore surrounded by smaller ones. At various locations on the colony these warts converge to form larger protuberances. This species also provides a habitat for burrowing barnacles (genus *Pyrgoma*), which cause other irregularities on the surface. It normally occurs in the shallows, near the reef crest generally at depths less than 5m.

Millepora dichotoma is a shallow-water species named after its characteristic branching growth pattern. It often forms large intricately interwoven fan-shaped colonies whose cylindrical, finger-shaped branches dichotomously divide at their tops. The orientation of the colonies is a good indication of prevailing current directions.

A closely related species, *Millepora tenera* occurs in the southern Red Sea and is widely distributed in the Indian Ocean. *Millepora exaesa* is recorded from the Gulf of Aqaba and from the Sudanese Red Sea.

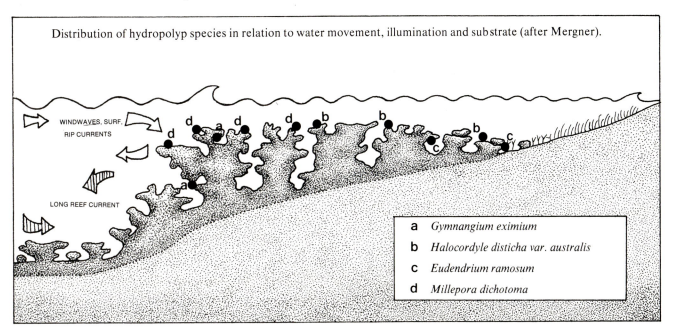

Distribution of hydropolyp species in relation to water movement, illumination and substrate (after Mergner).

WINDWAVES, SURF, RIP CURRENTS

LONG REEF CURRENT

a	*Gymnangium eximium*
b	*Halocordyle disticha var. australis*
c	*Eudendrium ramosum*
d	*Millepora dichotoma*

Above: Sea-fern: Hydroid: *Lytocarpus philip-pinus;* right: Stylasterid Hydrozoan; *Distichophora violacea.*

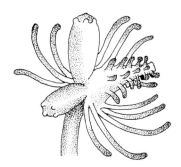

Halocordyle disticha var. *australis*

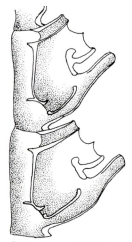

Lytocarpus philippinus

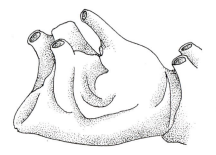

Gymnangium eximium (hydrotheca)

FAMILY: STYLASTERIDAE: Stylasterina Corals

Distichopora violacea forms generally small, delicately branched colonies whose unusual colour is a useful key to its identification. They are frequently dark violet with white tips to their branches. Others may be a pale, pinkish violet with darker tips. Their minute white tentacles may be seen along the narrow edges of their flattened branches. They are found in crevices among shallow-water corals, especially in areas where strong water movement occurs such as close to the reef-crest or on the reef-flat.

Ecology of Hydrozoans

Millepora dichotoma predominates in the areas of strong but rarely destructive surf such as occur along reef edges of the Central Red Sea and Gulf of Aqaba.

It possesses symbiotic zooxanthellae and requires strong sunlight and good water movement. *Millepora platyphylla*, which also lives in shallow reef zones where bright sunlight occurs, tends to prefer slightly less exposed locations such as the rear reef of lagoon fringing reefs or in the lagoon itself. *Millepora exaesa* is much less restricted in its choice of habitats and occurs in many situations from the crest down to at least 10m.

Like *M. dichotoma,* the cosmopolitan hydroid *Halocordyle disticha* also prefers well illuminated superficies close to the surface where turbulence occurs. The species is frequently seen growing on prominent surfaces at the reef edge, on the shallow reef-face, or on the reef-flat. It can also survive in areas of less water movement such as in shallow lagoons.

Another widespread tropical species is *Lytocarpus philippinus* which prefers clean water where wave caused oscillating currents provide a more or less continuous movement of water. It can survive at lower light intensities than *Halocordyle disticha* and has been found at quite considerable depths (e.g. trenches near Bab el Mandeb at 200m depth). It is found on open surfaces of the reef slope, usually at depths greater than five metres.

Eudendrium ramosum grows in areas of poor illumination and weak water movements such as occur in shaded crevices close to the shore, inside Red Sea fringing reefs. Such localities experience quite heavy sedimentation and this species is often found to be partially covered by fine silt.

A species favouring strong water movement but low light intensity is *Gymnangium eximium* which is often abundant in shaded crevices close to the reef crest or beneath overhanging rocks on the reef slope. As with *Millepora dichotoma,* the orientation of its fan-like colonies provides an indication of prevailing current direction.

TABLE I: MAJOR RED SEA HYDROPOLYPS*

FAMILY	SPECIES
Corynidae	*Sphaerocoryne bedoti*
Halocordylidae	*Halocordyle disticha* var. *australis* *Solanderia secunda, S. minima*
Hydractinidae	*Hydractinia echinata*
Eudendriidae	*Eudendrium deciduum, E. ramosum*
Haleciidae	*Halecium beanii*
Campanulariidae	*Campanularia gravieri, C. hemisphaerica* *C. latitheca, C. paulensis* *Laomedea (Obelia) bicuspidata, L. (Obelia) dichotoma*
Campanulinidae	*Cuspidella humilis*
Lafoeidae	*Filellum sp.* *Hebella parasitica, H. venusta* *Zygophylax armata*
Syntheciidae	*Synthecium elegans*
Sertulariidae	*Dynamena cornicina, D. quadridentata* *Diphasia digitalis, D. mutulata* *Thyroscyphus fruticosa* *Cnidoscyphus aequalis* *Sertularella campanulata* *S. mediterranea, S. natalensis* *S. polyzonias, S. ligulata, S. trigonostomata*
Plumulariidae	*Pycnotheca mirabilis* *Antennella secundaria* *Halopteris glutinosa* *Nemertesia ramosa* *Plumularia setacea, P. wasini* *Thecocarpus flexuosus* *Gymnangium eximium* *G. gracilicaulis, G. hians* *Aglaophenia latecarinata* *Lytocarpus balei, L. philippinus*

Sphaerocoryne bedoti.

Campanularia sp.

Hebella venusta
(hydrotheca)

Sertularella trigonostomata

*after Mergner & Wedler, 1977.

Scyphozoan jellyfish.

Class: Scyphozoa: Jellyfish

There are approximately fifteen species of jellyfish recorded from the Red Sea. Of these, *Aurelia aurita* and *Aurelia maldivensis* are probably the most abundant pelagic forms which, at certain times of year, form huge "rafts" which accumulate on the windward sides of reefs. Another species with which most divers or reef-walkers are familiar is *Cassiopea andromeda:* the Upside Down Jellyfish. This also lives in large aggregations and its normal mode of existence is resting on the sea-bed with mouth and tentacles uppermost while the exumbrella bell acts like a suction cup, clinging to the bottom.

The classification of recorded species is given in table II. Brief descriptive notes on each of the recorded species follow. The information is a condensation of that given by Kramp (1961) in his synopsis of medusae of the world.

Order: Cubomedusae

FAMILY: CARYBDEIDAE

Cubomedusae with four simple or tripartite interradial tentacles; four stomach pouches lacking diverticula.

Carybdea alata: 60 to 80mm high; approx. 50mm wide; exumbrella without warts; sensory niches enclosed by a pair of covering scales below and by a single one above; gastric filaments in crescentic areas extending horizontally at the corners of the stomach; tentacles simple.

FAMILY: NAUSITHOIDAE

Umbrella margin cleft into lappets; with a single mouth opening provided with simple lips; sense organs (rhopalia) and solid marginal tentacles arise from clefts between the lappets; circular, coronal furrow in the exumbrella, and peripheral to this a zone of gelatinous thickenings divided by radiating clefts.

Palephyra antiqua: 20mm wide; 8mm high; six to eight slender gastric filaments in each inter-radius; four interradial gonads crescent shaped with the horns recurved.

Moon jellyfish; *Aurelia aurita*

TABLE II: RED SEA JELLYFISH

ORDER: **CUBOMEDUSAE**

FAMILY: **CARYBDEIDAE**

Species: *Carybdea alata*

ORDER: **CORONATAE**

FAMILY: **NAUSITHOIDAE**

Species: *Palephyra antiqua*

ORDER: **SEMAEOSTOMEAE**

FAMILY: **PELAGIIDAE**

Species: *Pelagia noctiluca*

           Sanderia malayensis

FAMILY: **CYANEIDAE**

Species: *Cyanea capillata*

FAMILY: **ULMARIDAE**

Species: *Aurelia aurita*

           Aurelia maldivensis

ORDER: **RHIZOSTOMEAE**

FAMILY: **CASSIOPEIDAE**   **FAMILY:**   **THYSANOSTOMATIDAE**

Species: *Cassiopea andromeda*   Species: *Thysanostoma loriferum*

FAMILY: **CEPHEIDAE**   **FAMILY:**   **CATYLOSTYLIDAE**

Species: *Cephea cephea*   Species: *Crambionella orsini*

           Cephea octostyla   **FAMILY:**   **RHIZOSTOMATIDAE**

           Cotylorhiza erythraea   Species: *Rhopilema hispidum*

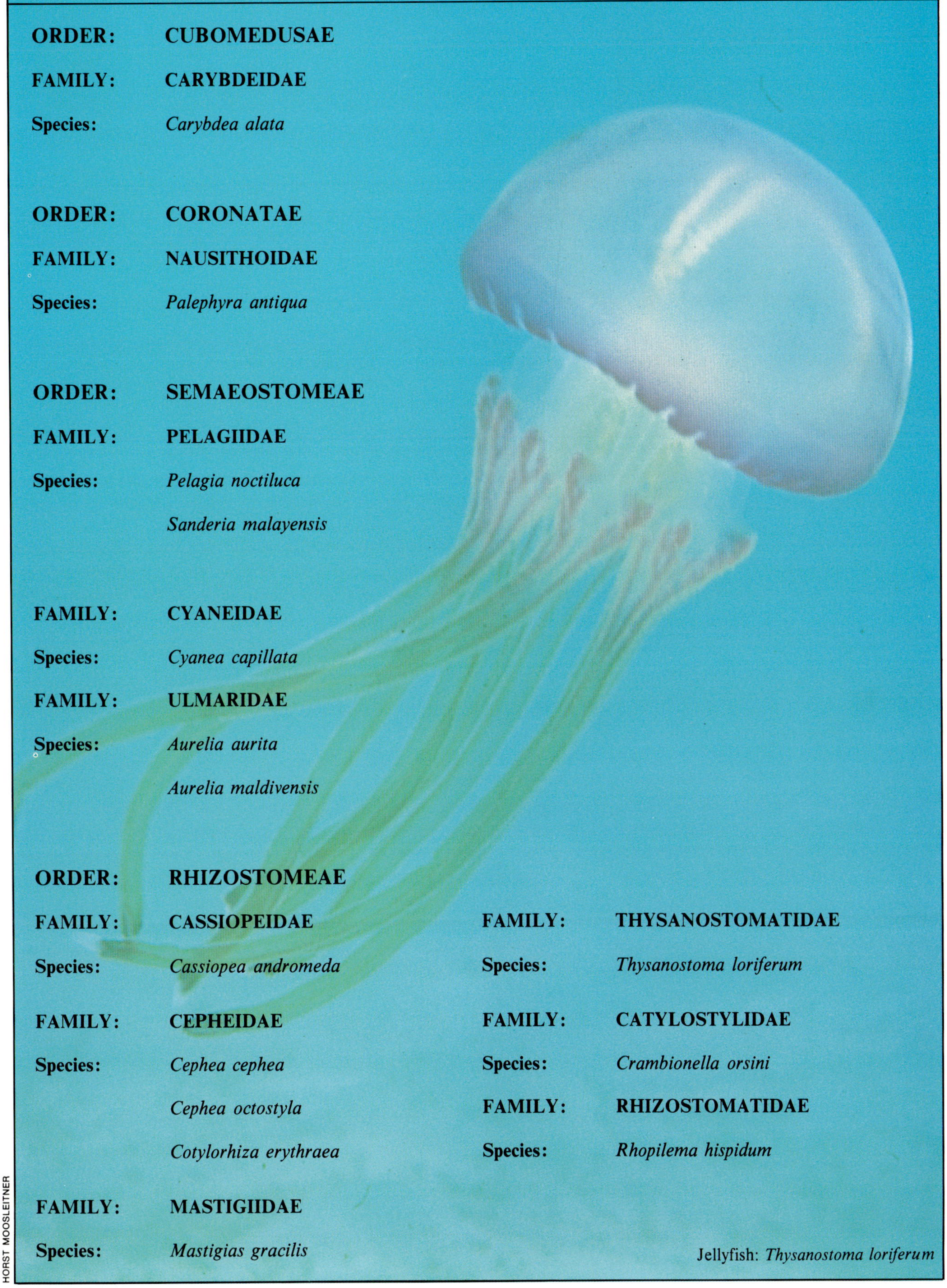

FAMILY: **MASTIGIIDAE**

Species: *Mastigias gracilis*   Jellyfish: *Thysanostoma loriferum*

HORST MOOSLEITNER

J. DAVID GEORGE

Jellyfish: *Pelagia noctiluca.*

Order: Semaeostomeae

FAMILY: PELAGIIDAE

Central stomach gives rise to completely separated, unbranched radiating pouches; without ring canal; tentacles arise from umbrella margin between clefts of the lappets; oral arms long, pointed, much folded.

Pelagia noctiluca: Up to approx. 65mm wide; colour very variable; sixteen marginal lappets; eight marginal sense organs; with eight tentacles alternating with marginal sense organs; sixteen similar radial stomach pouches, each terminating in a pair of unbranched canals entering the marginal lappets; exumbrella with numerous nematocyst warts, variable in size and number.

Sanderia malayensis: About 90mm wide; large nematocyst warts on central portion of exumbrella; four interradial heart-shaped genital ostia, bordered externally by twenty-four to thirty finger-shaped papillae; thirty-two cleft marginal lappets; sixteen marginal sense organs; sixteen tentacles alternate with marginal sense organs; thirty-two radial stomach pouches all alike.

FAMILY: CYANEIDAE

Central stomach gives rise to radiating pouches which in turn give rise to numerous branching, blind canals in the marginal lappets; no ring canal; gonads in complexly folded interradial sections of the wall of the subumbrella; tentacles arise from subumbrella at some distance from margin.

Cyanea capillata Diameter is up to 1m but usually much smaller in Red Sea specimens; rhopalar and tentacular stomach pouches completely separated; peripheral canals are somewhat curved with few or no anastomoses; colour reddish-brown or yellowish.

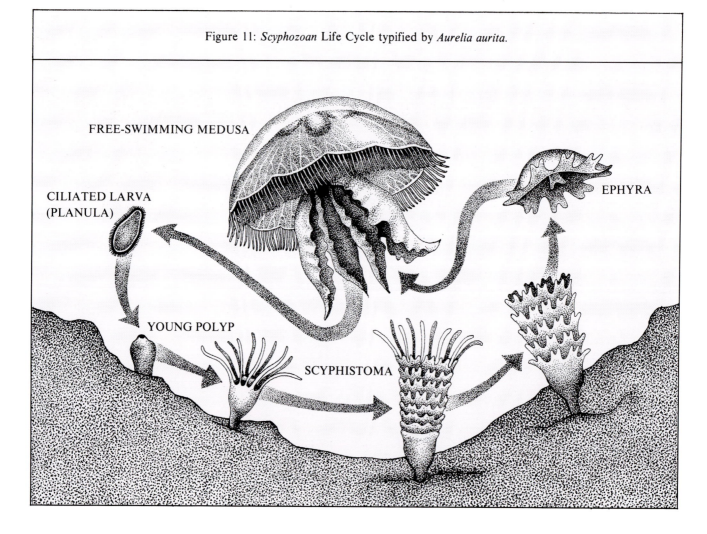

Figure 11: *Scyphozoan* Life Cycle typified by *Aurelia aurita.*

FREE-SWIMMING MEDUSA

CILIATED LARVA
(PLANULA)

EPHYRA

YOUNG POLYP

SCYPHISTOMA

FAMILY: ULMARIDAE

Aurelia maldivensis: Moon Jellyfish.
Up to 250mm wide; eight broad marginal lobes each with a slight central depression; eight rhopalar and adradial canals unbranched while the other canals are branched with occasional anastomosing; mouth arms large and curtain-like; lips complexly folded, with numerous short tentacles; colour violet.

Aurelia aurita: Moon Jellyfish
Up to approx. 400mm wide, with eight broad, simple marginal lappets; mouth arms as long as disc radius, thick and stiff with densely crenulated margins and numerous small tentacles; canal arrangement similar to *A. maldivensis;* sense organs in shallow clefts; colour variable. Feed on plankton by trapping food in streams of mucus which coat their upper and lower surfaces. Cilia carry mucus with trapped plankton towards the mouth. Typical Scyphozoan life-cycle; i.e. individuals male or female; sperm enter female via mouth and fertilisation is internal. Eggs develop into ciliated "planulae" larvae which emerge from mouth and can often be observed swarming on the undersurface of jellyfish. When planulae settle they form a "schistosoma" which eventually buds off to produce young medusae. This life-cycle is illustrated in figure 11.

Moon jellyfish: *Aurelia aurita*

Order: Rhizostomeae

Scyphomedusae with umbrella margin cleft into lappets; no marginal tentacles; without central mouth opening but with numerous mouths on eight fleshy, branched, arm-like appendages which arise from under the umbrella; with rhopalia between marginal clefts.

FAMILY: CASSIOPEIDAE

Cassiopea andromeda: 100-120mm wide; 20-30mm high; flat and disc shaped with a variable number of short, blunt, marginal lappets; mouth-arms wide, flat; four to six flat, short arborescent branches arise from each arm. Lives from extreme low water level to about 10m depth; in sheltered lagoons, mostly around sea-grass beds. Breeding occurs from April to August when they are most active and their stings are reputed to be more severe. They are sometimes eaten by turtles and by crabs (e.g. *Grapsus strigosus*).

FAMILY: CEPHEIDAE

Cephea cephea: 100-140mm wide; a large dome at apex covered with about thirty large, conical warts; eighty to ninety marginal lappets. Eight stout mouth-arms virtually coalesce in upper half but are forked and branched in lower half; in excess of 100mm long, tapering filaments. Characterised by very deep rhopalar clefts, long tapering mouth-arm filaments and brown colour.

Cephea octostyla: 90mm wide; 20mm high, exumbrella flat, rim vertical; exumbrella with a zone of numerous low warts leaving central portion free. Approximately seventy-two marginal lappets in groups of nine separated by slight indentations; eight bifurcated mouth-arms with numerous short filaments, and in middle region four to twelve long, tapering, warty filaments.

Cotylorhiza erythraea has numerous short radial canals in each octant; exumbrella with smooth central dome without warts; mouth arms with stalked suckers. Up to 90mm wide; four to six radial canals in each octant.

FAMILY: MASTIGIIDAE

Annular subumbrella muscles. Eight rhopalar radial canals with ring canal. Arm disc quadratic, with four primary canals. Short, pyramidal, three-winged mouth-arms; with filaments on the arm disc. Mouth arms terminate in a naked, club-shaped extremity; mouths along three edges of mouth-arms and also on their flat sides; numerous small clubs and filaments between frilled mouths.

Upsidedown jellyfish: *Cassiopea* sp., showing greenish-brown colouration caused by symbiotic algae.

Mastigias gracilis: 35mm wide, thin at margin but very thick at apex, exumbrella with irregularly arranged clusters of small warts; margin irregularly lobed with five to ten lappets in each octant; arm disc with particularly long filaments; mouth-arms not as long a disc radius and the lower three-winged section is three to four times as long as the upper portion; each arm has a short rounded terminal knob and short gelatinous knobs between frilled mouths; six to seven canal roots in each octant.

FAMILY: THYSANOSTOMATIDAE

Annular subumbrellar muscles; eight rhopalar radial canals; ring canal present; arm disc quadratic with four primary canals; long, narrow, lash-like mouth arms, triangular or three-winged in cross-section, no clubs or filaments.

Thysanostoma loriferum: Up to 200mm wide; exumbrellar smooth or finely granulated; six to eight velar lappets in each octant, united by a membrane; mouth arms terminate in a short, oval, naked knob; fine meshed intra-circular canal system with up to thirty canal roots in each octant.

FAMILY: CATOSTYLIDAE

Mouth-arms three-winged. Network of anastomosing canals issue from permanent ring canal; subumbrellar muscles annular; no scapulets; not always sixteen radial canals; eight rhopalar canals extend to umbrellar margin; eight inter-rhopalar canals extend only to ring canal; mouth arms are pyramidal.

Crambionella orsini: 100-200mm wide; plump; massive; hard and cartilaginous; smooth; sixteen small sharp velar lappets separated by furrows which extend up along the surface of the exumbrella; mouth-arms approximately as long as bell radius; proximal portion short, only one third length of distal portion.

FAMILY: RHIZOSTOMATIDAE

Mouth-arms coalesced proximially and distal portion three-winged; usually with a terminal club; network of anastomosing canals issue from primary ring canal; subumbrellar muscles annular; eight pairs of scapulets on upper arms; sixteen radial canals all extend to umbrella margin; four subgenital cavities; no primary mouth opening.

Rhopilema hispidum 250-340mm wide; exumbrella with numerous small, sharp pointed, conical projections; eight velar lappets per octant; mouth arms terminate in a large club shaped appendage with a faceted, swollen end; other club shaped appendages on the three wings between the mouths.

Opposite: Cerianthid
Below: Jellyfish: *Cotylorhiza erythraea;* right: *Cassiopea* sp.

F. JACK JACKSON

J. DAVID GEORGE

F. JACK JACKSON

Class: Ceriantipatharia
Black Corals, Thorny Corals and Cerianthids

Order: Antipatharia

FAMILY: ANTIPATHIDAE
Antipathes dichotoma: Black Coral

Black coral has been known to, and valued by, local people for a considerable period. Before the advent of diving it was hauled up occasionally on boat anchors and colonies were also found in some relatively shallow locations such as inside natural harbours along the coast. The axial skeleton of black corals is formed by a brown or black extremely hard proteinaceous material which, when cut and polished, is used for making jewellery and prayer beads.

The large black coral "trees" which are found on many Red Sea reefs, usually in relatively deep water at around 50m, or else on shaded overhanging reef-faces in shallower water, are likely to be *Antipathes dichotoma* which is the largest black coral species in the Red Sea.

Cirrhipathes anguina: Whip Coral

Whip corals are also abundant in deep water around Red Sea reefs. Their colonies consist of long, thin "stalks" which are frequently bent or coiled towards their extremities.

Order: Ceriantharia: Cerianthids

FAMILY: CERIANTHIDAE

These anemone-like Coelenterates are usually found in mud or sand around the base of coral-reefs. They live in mucus tubes which are buried in the sediment. At the approach of any disturbance they rapidly withdraw into their tubes. One feature which distinguishes them from true anemones is the fact that they have two types of tentacles; shorter ones surrounding the mouth and longer ones outside these.

Pachycerianthus mana is a night feeding cerianthid which lives at moderate depths (e.g. 1-20m), usually in fine sand or silt. During daytime they are usually hidden in their mucus sheaths but at night they expand and stretch their long tentacles into the water column in order to catch fish and plankton. In many instances, however, the presence of *P. mana* can be detected as a result of the slight protrusion of its mucus tubes above the sea bed. Observations on the behaviour of this species have shown that, during feeding, each tentacle responds to stimuli, independent of the other tentacles. Thus, when feeding on small prey items such as copepods, a single tentacle can paralyse, adhere and transport the food to its mouth whilst when taking larger items such as fish, several tentacles may work together. Fishelson (1970) reports finding *Phoronis* sp. living in the mucus tube of *P. mana* collected at 12m in the Gulf of Aqaba.

Cerianthid

Cerianthid: *Pachycerianthus mana*

Opposite: Black coral: Antipatharian; left: Whip coral: *Cirripathes anguina* with *Xenia* sp.

Class: Alcyonaria: Octocorals

These have eight feather-like ("pinnate") tentacles. They are colonial forms which frequently have internal skeletons composed of calcareous or horny material.

Order: Stolonifera

FAMILY: TUBIPORIDAE

Distinctive octocorals with very long polyps and complete fusion of spicules to form a solid hard mass. Flat horizontal platforms or laminae connect close set polyps and from these platforms, new polyps may arise.

Tubipora musica: Organ-pipe Coral
This species occurs in very shallow water on the reef-platform, usually towards the outer reef-crest where it grows in crevices protected from the most vigorous wave-action. Its common name results from the arrangement of its deep red skeleton which is formed by parallel rows of tubes which are bound together by transverse septa. In life the skeleton is usually obscured by its green-grey polyps each of which has eight radiating feathery tentacles. When touched the polyps retract to reveal the skeleton. It is a coral which, as a result of its wave-washed habitat, is frequently dislodged during storms and fragments are thus cast ashore where they form attractive collector's items.

FAMILY: CLAVULARIIDAE

Clavularia hamra. This octocoral (figure 12) has a fairly wide distribution in the Red Sea and occurs in particular around the base of coral pinnacles or reef-faces in relatively sheltered locations on a variety of substrates including *Sargassum* weed, sponges and dead corals. Expanded colonies are chocolate brown with light yellow or greenish yellow spots. When contracted their colonies are a dirty purple colour. Polyps arise from a ribbon-like stolon and when fully expanded they are up to 12mm long. Their colonies are strengthened by numerous warty spicules. It should not be confused with *Xenia* which it superficially resembles. Like *Xenia*, it appears to depend on its zooxanthellae to provide its nutritive requirements. Close examination of colonies of this soft coral are likely to reveal a look-alike nudibranch: *Pleuroleura striata* which feeds on this Alcyonarian.

Clavularia pulchra is a closely related species which occurs on pearl shells, other molluscs and stones in sheltered harbour areas such as Khor Dongola along the Sudanese coast.

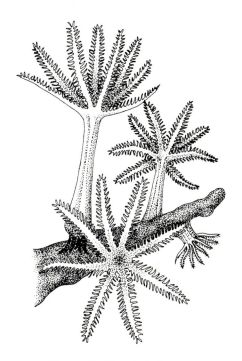

Figure 12: *Clavularia hamra.*

Opposite and below: Organ-pipe coral: *Tubipora musica.*

PETER VINE

Soft coral: *Sinularia polydactyla*

Order: Alcyonacea: Soft Corals

Soft corals are characterized by the fact that their colonies are composed of fleshy, sometimes lobed masses from which extend numerous generally retractable polyps. They are particularly abundant on Red Sea reefs where they frequently compete for space with the true stony corals. Some species can form huge colonies which may extend to cover as much as ten metres of horizontal reef-face. Their great mass results primarily from a huge thickening of the gelatinous layer in the wall of polyps. A certain amount of support is provided by a loose network of embedded spicules whose structure is used as a major taxonomic characteristic to separate closely related species. As with other members of the class, the soft corals have eight pinnate tentacles. A list of species is provided in Table III.

FAMILY: ALCYONIIDAE

Cladiella pachyclados is grey-white with brownish tentacles. It forms clumps in which closely packed finger-like projections bear polyps.

Sinularia gardineri forms flat, low growing colonies in which the capitulum consists of a main lobe and up to six secondary lobes arising from this. The general colour is grey-cream to cream. The stalk is short and irregular, sloping outwards as it meets with the capitulum.

Sinularia leptoclados is greenish-grey with slender, rather knotted finger-like branches and branchlets up to 1.4cms long, with a diameter of 4mm arising from near the base of the stem.

Sinularia polydactyla forms large densely lobed colonies with finger-like projections varying in length; brown or grey or greenish-cream in colour. As in all soft corals the structure of spicules is a taxonomically important species characteristic. This species possesses long pointed spindles and club-shaped spicules in which there are several warty projections.

Table III
RED SEA SOFT CORALS*
Parethropodium fulvum
Cladiella pachyclados
Lobophytum pauciflorum
Sarcophyton ehrenbergi, S. glaucum, S. trocheliophorum
Sinularia grayi, S. gardineri, S. querciformis, S. leptoclados, S. polydactyla
Litophyton arboreum
Dendronephthya hemprichi, D. klunzingeri
Stereonephthya cundabiluensis
Umbellulifera oreni
Siphonogorgia mirabilis, S. fragilis
Xenia crassa, X. membranacea, X. blumi, X. macrospiculata, X. impulsatilla. X. hicksoni, X. obscuronata, X. mayi, X. umbellata, X. sansibarina
Heteroxenia fuscescens
Anthelia glauca

*after Verseveldt, 1965

Sarcophyton trocheliophorum is an abundant species found in sheltered habitats such as Suakin harbour, where it grows in close association with hard corals. When tentacles are retracted it appears greenish-brown, but when they are extended the colony may appear, from above, to be white. A fairly stout, short, trunk carries a spreading and convoluted "cap" on which are situated numerous polyps. When seen underwater, it appears to consist of a number of closely inter-locked fleshy lobes.

Lithophyton arboreum is another common species of soft coral which occurs on Red Sea reefs, frequently in silty habitats. It has a tree-like branching form with polyps born on thick arborescent extensions from the main "trunk". Olive green and with branches rising 50 to 60 cms above the substrate. Provides a refuge for small damsel fish. Best observed in the late afternoon when its polyps tend to be expanded.

Lobophytum pauciflorum possesses a low sterile stalk (with five longitudinal striations) on which is born a broad capitulum. The latter is covered with a number of finger-like lobes. Well separated polyps cover the lobes but they are more closely packed together at the tips. It is common on shallow reefs where it forms large grey-yellow colonies with fleshy lobed extensions patterned by its yellow polyps.

Parerythropodium fulvum forms thin, yellow sheets on dead coral, especially on the exposed reef-face.

A young colony of *Sarcophyton* sp.

Sarcophyton sp.

Sarcophyton trocheliophorum

Sarcophyton trocheliophorum

Sarcophyton trocheliophorum

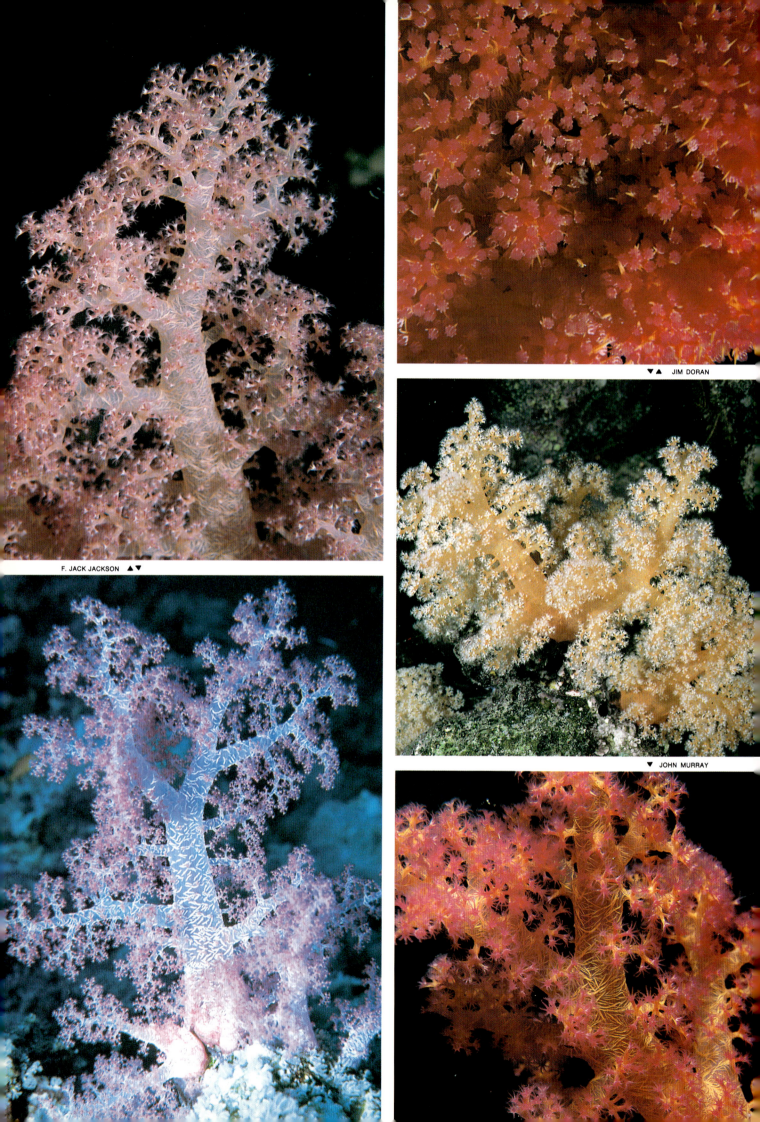

F. JACK JACKSON ▲▼

▼▲ JIM DORAN

▼ JOHN MURRAY

JOHN MURRAY

FAMILY: NEPHTHEIDAE

Dendronephthya hemprichi is one of the most beautiful species. Unlike most soft corals which tend to be rather dull in colour, these are bright pink or orange with a translucent form. They have a very short stem which may be completely absent. The branching lobes tend to splay out in a single plane (unlike those of *D. klunzingeri* which form large bunches). Polyps are grouped together and arise from thick fleshy translucent branches in which are clearly visible spicules. In this species the bundles contain from six to eleven or more polyps, whereas those of *D. klunzingeri* have only one to three polyps in each bundle. It grows on overhanging surfaces in relatively shallow water and in more open areas of the reef at greater depths.

Dendronephthya klunzingeri is closely related to and frequently confused with *D. hemprichi*. Positive separation of the two species is based on details of their calcareous spicules. The polyps are as pink as cherry blossoms while its main branches are translucent and colourless. It lives in shaded locations where there is a moderate water exchange; usually attached to the undersides of coral knolls on deeper terraces or reef-faces. Colonies tend to contract in daytime but may reach 1m or more in height when fully expanded at night.

Opposite and above: selection of the soft coral, *Dendronephthya* sp. showing the spectacular colour range.

F. JACK JACKSON

FAMILY: XENIIDAE

Xenia umbellata. The genus *Xenia* is particularly successful in colonizing the skeletal remains of hard corals and indeed, any other suitable surfaces on the reef where it can establish a "foothold". Its colour is somewhat variable, with its eight pinnate tentacles ranging from light grey to dark brown. These often pulsate in a rhythmic opening and closing motion, thus helping to create a current of water across their surface.

It is found in many shallow water situations such as, for example, in Jeddah harbour on concrete structures and dead coral and at Obhour creek (Saudi Arabia) on the pilings of various jetties.

Polyps of Xeniid soft corals cannot be retracted. Despite the fact that their tentacles open and close in what appears to be a feeding pattern, they are believed to depend almost entirely on their contained symbiotic algae (Zooxanthellae) for their food requirements. At least ten species of *Xenia* have been recorded from the Red Sea.

Heteroxenia fuscescens polyps have a tall (5-8 cm long) column which does not usually branch. Colour pale grey or pale brown. Tentacles contract synchronously and rhythmically at a much more rapid rate than those of *Xenia umbellata* (i.e. 30-40/min in *H. fuscescens*). This coordinated tentacular motion must increase current flow over the colonies and have advantages for respiration, if not also for nutrition.

Anthelia glauca. Whereas the polyps of *Xenia* species arise from stout trunks, *Anthelia* polyps branch from their point of attachment to the substrate. This form is less common than *Xenia* and *Heteroxenia* and tends to occur in somewhat deeper water.

Opposite: close-ups of Xeniid soft corals; above: *Xenia* cf *elongata.*

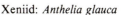

Xeniid: *Anthelia glauca*

J. DAVID GEORGE

Order: Gorgonacea: Horny Corals
This order includes the sea-fans and sea-whips.

FAMILY: SUBERGORGIIDAE
Subergorgia hicksoni. The original type material was collected from the piles of a steamer pier at Ghardaqa. It seems to prefer well-shaded habitats where there is relatively strong water movement. Thus, it may be found in fairly deep water on the exposed side of offshore reefs where wave influence is felt. It forms large upright fans (up to a metre or more in height) which are attached to the reef surface or to rocks lodged in the sand.

FAMILY: MELITODIDAE
Acabaria erythraea is a delicately branched red gorgonian which is sometimes found alongside the more abundant shallow-water species *A. pulchra,* in crevices near the reef-edge where shaded micro-habitats exist in conjunction with strong water currents. It also occurs deeper down the reef-face where it provides a habitat for the hawkfish, *Oxycirrhites typus.*

Acabaria biserialis is an irregularly branching gorgonian occurring in shaded localities on the reef-face, usually in the 10 to 20m depth zone. On shaded harbour pilings it may be abundant from as little as 2m deep where it replaces *Acabaria pulchra,* which requires more turbulent water movement than *A. biserialis.* Fans may reach 30cms in height and they grow so that they face into prevailing long-reef currents, thus maximizing the effectiveness of their plankton feeding network.

Acabaria pulchra is a small branching gorgonian with rigid colonies up to about 6cms in height. It occurs intertidally and in shallow water, always in well-shaded habitats, often attached to the undersides of boulders or in crevices. It is common throughout the Red Sea and is frequently present on harbour pilings, just below low-water level. Although it requires a well-shaded habitat (maximum 2% of surface illlumination), it also prefers strong water movement such as that created by wave surge.

Opposite and above: Gorgonian Sea-Fans

Gorgonian: *Acabaria erythraea*

Gorgonian: *Acabaria biserialis*

Gorgonian

Rumphella sp. soft coral in strong current.

Sea-pen: *Scytaliopsis ghardagensis*
Right: Close-up of Gorgonian; opposite;
Whip coral: *Juncella juncea.*

FAMILY: MELITHAEIDAE

Melithaea squamata Large orange fans which protrude from the reef-face in deep water may belong to this species.

FAMILY: PLEXAURIDAE

Rumphella sp. The accompanying photograph illustrates a specimen of gorgonian from the Sudanese Red Sea which is tentatively identified as *Rumphella aggregata*, an Indo-Pacific species which is found in relatively shallow water, in areas of high light intensity and strong water currents. The photograph shows a colony with its elongate branches bent over by a powerful current.

FAMILY: ELLISELLIDAE

Junceella juncea is a whip coral growing in deep water where it generally lives attached to hard substrates. The ends of their "whips" are usually somewhat curved.

FAMILY: GORGONELLIDAE

Gorgonella maris-rubri. This large bright orange gorgonian grows in fairly deep water, where it forms upright colonies of occasionally bifurcating, long narrow "arms". The type specimen came from "40 fathoms" in the Gulf of Suez but it occurs throughout the Red Sea at comparable depths.

Order: Pennatulacea: Sea-pens

Sea-pens are colonial octocorals which have a central "stalk" formed from a modified polyp and a series of short fleshy polyps arising from this. They are generally found living in soft sediments and may be encountered more frequently on night dives (when their colonies are normally extended) than in daytime. Not all species are able to retract into the mud however.

Scytaliopsis ghardagensis is found in soft sediments around Red Sea reefs (see photograph).

HORST MOOSLEITNER

Anemone: *Boloceroides mcmurrichi*

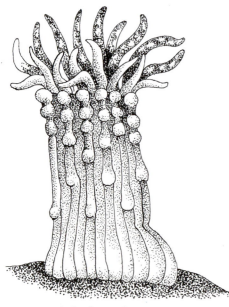

Anthopleura elatensis

Below: Tentacles of *Stoichactis* anemone; opposite: original illustrations of Red Sea anemones by Klunzinger (1877).

F. JACK JACKSON

Class: Zoantharia

In this class polyps have six (or multiples of six) simple tentacles. It includes solitary and colonial forms such as sea anemones, colonial anemones and corals.

Order: Actiniaria: Sea Anemones

FAMILY: BOLOCEROIDIDAE

Boloceroides mcmurrichi is a somewhat unusual small anemone which measures 1-3cms in diameter and is able to swim by beating its tentacles. Since it frequently reproduces by asexual budding it may be found in small aggregations. It tends to occur in fairly sheltered locations such as among mangrove roots or attached to spines of the black-spined sea-urchin, *Diadema setosum*.

FAMILY: ACTINIIDAE

Anthopleura elatensis occur on sandy, shallow areas, partly covered by sand. Often on undersides of stones in mid-tidal (northern areas) along with *Planaxis* and *Tetraclita*. Body mostly enclosed in a mucus sheath to which sand grains adhere. Colour: pale brown with darker patches or dots on upper part of column. Tentacles with pale greenish lines. When found in day, or at low-tide, they are normally contracted whereas at night, and high tide, they expand and feed on plankton by means of their long tentacles. Frequently multiplies by fission.

FAMILY: ALICIIDAE

Alicia mirabilis. Both *A. mirabilis* and *A. zanzibarica* are recorded from the Red Sea. This is a nocturnal feeding anemone which extends its long stinging tentacles into the current during darkness. When drifting plankton come in touch with them the tentacles immediately contract and bend towards the mouth. It has prominent "warts" formed by groups of stinging cells on the column. In daytime the column is withdrawn into a tube but at night it is extended out of the soft substrate in which it lives and the stinging cells thus protect the column from nocturnal predators.

Triactis producta, which possesses powerful stinging nematocysts, occurs on the claws of the boxer crab *Lybia leptochelis* as well as in a free-living mode. The crab occurs from the lowest inter-tidal zone to a depth of about 1.5m and is usually found under stones or corals. The anemones which almost invariably occur on their chelae are usually pinkish and they may have lateral buds. According to Fishelson (1970) it is very difficult, if not impossible, to keep the crabs alive for long in an aquarium without their symbiont anemones. The anemone itself has a much wider distribution than that provided by its association with boxer crabs. It occurs on many surfaces, often in shaded crevices or among the basal branches of *Millepora dichotoma*. These adult forms are brownish red with pinkish vesicles around the oral-disc.

FAMILY: STOICHACTIDAE

Stoichactis gigas is the large anemone which is frequently seen at moderate depths along the reef face. It has greenish tentacles. It is usually accompanied by the common clownfish *Amphiprion bicinctus* and sometimes also with juvenile damsel fish *Dascyllus trimaculata*. The fish use the anemone for refuge from predators while the anemone is able to feed on fragments of food dropped by the fish. The anemone also benefits by a certain amount of cleaning activity carried out by the clownfish. In addition, the wafting of the fish's tails across the anemone aid it in creating a respiratory current.

It has been shown that the clownfish avoid being stung by the anemone as a result of a slimy mucus covering on the clownfish. It is not clear whether this mucus is originally secreted by the fish or the anemone.

Stoichactis tapetum occurs in coral crevices where it lives partially embedded in coralline debris forming the sediment. The column itself spreads over the surface of the sediment and the oral disc is covered with very short, simple tentacles. The general colour is brown-red to brown-yellow with a pattern of small red spots.

Ehrenberg del.

W. A. Meyn lith.

1. Heterodactyla Hemprichi Ehrb. nat.Gr. 2. Actineria Hemprichi Ehrb. ½ ⅔ nat.Gr. 3. Phymanthus loligo Ehrb. nat.Grösse.
4. Paractis pulchella Ehrb. 2 fach vergrössert. 5. Phellia decora Ehrb. nat.Gr.

Clown-fish: *Amphiprion bicinctus* with *Stoichactis* anemone.

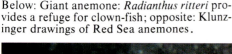

Below: Giant anemone: *Radianthus ritteri* provides a refuge for clown-fish; opposite: Klunzinger drawings of Red Sea anemones.

Radianthus ritteri is another large anemone which is often found in shallow water, often close to the reef edge where there may be relatively strong turbulence. It settles on hard substrata such as at the base of massive corals. The outer column of the anemone is a beautiful shade of lilac, while the numerous tentacles are a mixture of brown, orange and beige.

Radianthus koseirensis occurs in shallow water, attached to dead coral either among patches of live corals or sometimes on sandy areas, where its rocky substrate is buried under the sediment. Suction cups are arranged in rows covering the upper portion of the body, which is frequently covered by sand and small stones. The tentacles are all of similar length. Both this and *R. ritteri* frequently have anemone fish associated with them. It grows down to 15m depth on the fore-reef of Aqaba.

Gyrostoma helianthus has a low, broad, orange or dark pink column and many short pale tentacles arranged in concentric rows, with inner ones being longest. It was found to be present on the fore-reef at Aqaba from the surface to 40m depth, with its maximum abundance in the 5-15m depth range. Disc up to half a metre in diameter and anemone can reach about 75cms in height. Usually has clown-fish: *Amphiprion bicinctus* in attendance.

Gyrostoma quadricolor is frequently found lodged among dead corals such as, for example, the loose coralline fragments which occur on shallow lagoons where damsel fish establish, and fiercely defend, their territories. In its partially withdrawn state it has typically bulbous light green-grey translucent tentacles which are bunched together as they protrude from the crevice which houses the anemone. It has a wide depth distribution, occuring from the surface to at least 40m depth.

FAMILY: PHYMANTHIDAE
Phymanthus loligo is found, embedded in sediments, among corals in shallow depths. It is a small anemone (column length approx. 4 cms); reddish brown in colour and with lateral outgrowths on the tentacles.

Ehrenberg del.   W. A. Meyn lith.

1. Bunodes crispus Ehrb. ⅓ nat. Gr. 2. Thalassianthus aster Leuck. nat. Gr. 3. Rhodactis rhodostoma Ehrb. ½ nat. Gr.
4. Paractis adhaerens Ehrb. ½ nat. Gr. 5. Paractis Hemprichi Klz. nat Gr.   6. Paractis erythrosoma Ehrb. ⅔ nat. Gr.
7. Paractis erythräa Ehrb. nat. Gr.   8. Paractis olivacea Ehrb. nat. Gr.

Clown-fish *Amphiprion bicinctus* with anemone: *Gyrostoma helianthus.*

FAMILY: THALASSIANTHIDAE

Heterodactyla hemprichi live attached in crevices and under coral boulders. They have flat oral discs with multi-branched brown tentacles (up to 20mm long) surrounded by longer, thicker marginal tentacles which branch laterally and have globular violet nematophores between them.

Cryptodendrum adhesivum has a smooth body and many short, sticky tentacles and is sometimes confused with *Stoichactis gigas*. It has a white body with reddish or pinkish spots, while the foot is often reddish with yellow spots. Tentacles are brown to red with white tips. It may be accompanied by the small shrimp *Periclimenes brevicarpalis* and lives in shallow reef-areas, frequently lodged between corals.

Thallassianthus aster forms colonies in very shallow water. Reproduces by budding. Often occurs on concrete pilings, immediately below the *Tetraclita/Planaxis* zone; forming a dense "matting" just below water level. Colour varies: light-pink to brown or blue-green.

FAMILY: HORMATHIIDAE

Calliactis polypus is found on shells inhabited by the hermit crab *Dardanus tinctor* (see page 108). It has a pale column with brown dots and purple lines. The disc is grey-violet with yellow dots and the oral disc is somewhat pinkish. Tentacles which are about 1-5cms long are grey, violet, white or spotted with yellow and ringed. It is about 3cms long and around 3-4cms broad.

Zoanthid: *Palythoa tuberculosa*

Order: Coralliomorpharia

FAMILY: ACTINODISCIDAE

Actinodiscus nummiformis sometimes forms loose colonies in fairly shallow water, often on the underhanging surfaces of crevices. It reproduces by budding. Colour is variable, bluish-green, red or brown.

Rhodactis rhodostoma is usually brown in colour with a dense covering of short, papillous tentacles. Lives on dead corals in shallows.

Order: Zoanthiniaria

These are mostly colonial forms whose polyps are superficially similar to those of small anemones.

FAMILY: ZOANTHIDAE

Palythoa tuberculosa (see photograph) occurs in shallow water, usually among dead corals. When withdrawn the polyps give the appearance of a stony coral. It is a highly toxic sea anemone and may render certain fish which consume it poisonous to eat.

Zoanthus bertholetti forms mat-like colonies, usually on sand or sand-covered rocks. The polyps are approximately 5mm high and 3mm wide, and the base of the column is blue-grey while the upper portion is bright red. The oral disc is granulated. It is quite a common anemone on stones in shallow water.

Zoanthid soft coral: *Palythoa* sp.

JIM DORAN

4. COELENTERATES II

Stony Corals

Corals have been of prime interest to many marine scientists who have worked in the Red Sea. Reports on their taxonomy date back to 1738, when Thomas Shaw published the results of a journey through the Middle East which included visits to Egypt and the Sinai peninsula. His descriptions were inadequate for accurate species designation however, and the earliest truly scientific collection of Red Sea corals was made by Peter Forskål, who was a biologist on the Danish expedition better known as the "Arabia Felix Expedition" in 1762 and 1763. Although he died during the trip, in the mountains of Yemen, part of his collection eventually reached Copenhagen and his work was posthumously published, thanks to the single survivor from that epic adventure — Carl Niebuhr. Twenty-six coral species were described by Forskål. Since then important studies have been carried out by Ehrenberg (1834); Klunzinger (1877, 1879-80); Marenzeller (1906); Gardiner (1909); Crossland (1935, 1935a, 1938, 1952); Scheer (1962, 1964, 1967, 1971, 1983); Fishelson (1973a); Loya (1974, 1976); Loya and Slobodkin (1971); Mergner (1971) and Head (1980), and by other authors too numerous to discuss in detail. At the time of writing this book, several recent studies provide a useful review of taxonomy and ecology of Red Sea corals. A brief list of such references would include at least the following:

Youssef Loya and L.B. Slobodkin (1971)
The Coral Reefs of Eilat (Gulf of Eilat, Red Sea)
Symp. Zool. Soc. Lond., 28: 117-139.

Hans Mergner (1971)
Structure, ecology and zonation of Red Sea Reefs.
Symp. Zool. Soc. Lond., 28: 141-161.

Brian Rosen (1971)
The distribution of Reef Coral genera in the Indian Ocean
Symp. Zool. Soc. Lond., 28: 263-299.

George Scheer (1971)
Coral reefs and coral genera in the Red Sea and Indian Ocean
Symp. Zool. Soc. Lond., 28: 329-367.

George Scheer and C.S.G. Pillai (1983)
Report on the stony corals from the Red Sea
Zoologica, pp. 1-198;41 plates.

Stephen Head (1980)
The ecology of corals in the Sudanese Red Sea
D.Phil. Thesis, Cambridge University.

Elisabeth Wood (1983)
Corals of the World. TFH Publications. 256pp.

Scheer and Pillai (1983) review the literature and research carried out to date and provide a comprehensive study of Red Sea corals from a taxonomic viewpoint. Their study was based upon 2,074 collected specimens comprising 194 species belonging to 70 different genera. This massive collection also covers a wide geographic area including the Gulf of Aqaba, Gulf of Suez, northern, central and southern areas of the Red Sea. They were drawn from nineteen different sources and include

F. JACK JACKSON

Opposite: submerged reef top: above: close-up of *Favites* sp. stony coral.

the results of various collections made between 1914 and 1981. Of the 194 species identified, 161 (51 genera) belong to hermatypic (reef-building) corals and 33 (19 genera) are ahermatypic.

It is interesting to compare the above, extremely comprehensive study with that made from a much more restricted region of the Sudanese Red Sea by Stephen Head (1980). He studied 1200 specimens, the vast majority of which were methodically collected during transect studies around a single patch reef. From this he identified 132 species of calcareous cnidarians belonging to 63 different genera or subgenera, reflecting the remarkable diversity of forms which occur in the Red Sea.

To a certain extent, the number of coral species recorded from an area may reflect the amount of effort which has been expended on collecting and identifying the species which are present. However, based upon the many collections which have been investigated from different regions, a general picture is beginning to emerge. This suggests a maximum diversity of species and coral genera in the mid-western area, and least variety in the Gulf of Suez in the north and around the Straits of Bab el Mandeb in the south. This is illustrated in figure 13.

Stagshorn coral: *Acropora* sp.

Stylophora pistillata

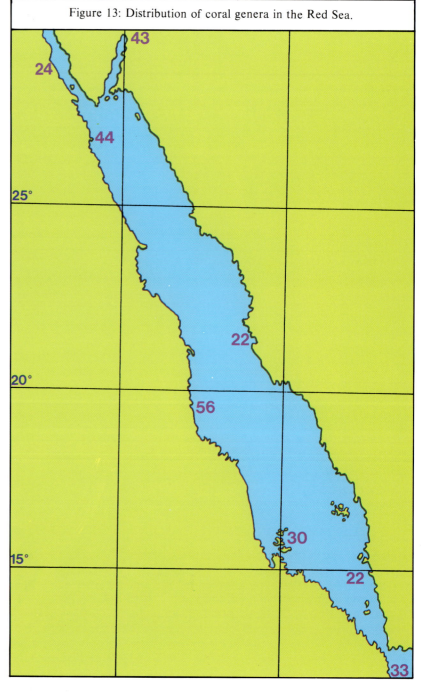

Figure 13: Distribution of coral genera in the Red Sea.

Ecology of Corals*

It is quite clear from the above data that corals flourish in the Red Sea and that the reefs they have created exhibit a rich variety of species. This prolific development may be partially explained by the near optimum environmental conditions which exist. The most favoured temperature for reef-building corals appears to be 25°C to 29°C. Favoured salinities are 34-36%.

Most reef-building corals require adequate light intensities for photosynthesis by their contained zooxanthellae (symbiotic algae). The true role of these single-celled algae in the nutrition of corals is still a subject of debate and continuing research. While some workers have claimed that corals gain most of their nutrition from feeding on zooplankton, others have demonstrated growth of corals in plankton free environments when they have received adequate light. It is clear that light is essential for the calcification process in reef-building corals, and it is therefore not surprising that such corals are only found growing at depths where suitable light penetrates the sea-water.

Reef edge of fringing reef at Rabigh, Saudi Arabia.

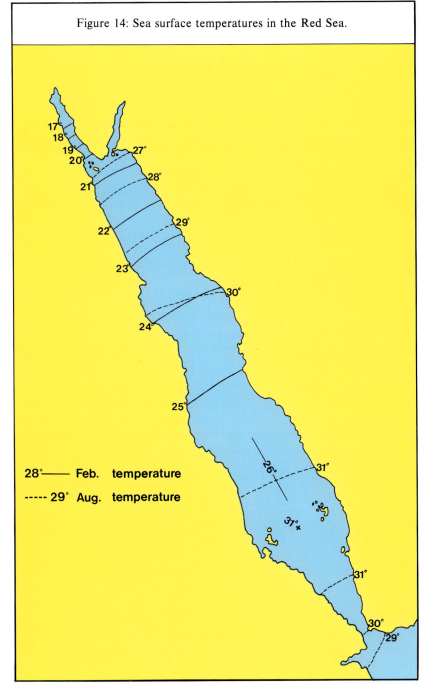

Figure 14: Sea surface temperatures in the Red Sea.

28°—— Feb. temperature

----- 29° Aug. temperature

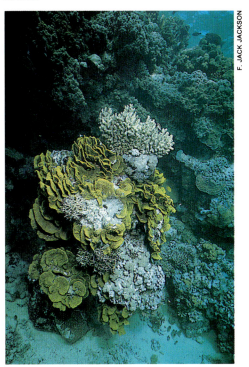

Mixed stony corals: *Acropora* sp., *Seriatopora hystrix,* and yellow, leafy *Turbinaria mesenterina.*

*The ecology of Red Sea corals is discussed in considerable detail by Stephen Head (1980). The following comments are based partly upon his observations and partly upon personal experience.

It has been suggested by several workers that while some corals depend mainly on zooxanthellae and their photosynthetic process for their nutrition, other species may depend mainly upon trapping plankton and depend less upon light energy. An interesting postulation has been made with regard to the different growth forms which corals at opposite ends of the spectrum are likely to adopt. The light dependent shallow water corals have branching or flat, spreading forms with small corallites and thus a high surface area to skeletal volume ratio, while mainly planktivorous corals might be expected to have large polyps with tentacles active in food capture and large robust skeletons. In some corals which appear to be active plankton feeders the polyps are small and they use ciliary feeding currents. Light loving corals which tend to be fast growers can smother many other organisms in shallow water, whereas they are much less successful in deeper water where the latter planktivorous group tends to come into its own. This general scheme does shed some light upon the vertical distribution of corals but it also throws up numerous apparent anomalies. It is to be expected that while some species may comply closely with either end of the spectrum, others will have intermediate characteristics and some may adapt different growth forms according to their ambient conditions.

In shallow areas of the Red Sea where sea-temperatures are close to their optimum range for coral growth, the effects of water movement may favour certain species and dictate against others. The most obvious example of this can be observed in relation to exposure to wave-action. The central Red Sea is unlike many coral areas, however, in that storms with wind force greater than Force 7 are unusual. Corals growing on reef-faces exposed to prevailing winds are constantly battered by waves and are adapted to withstand these forces, whereas those growing on the sheltered sides of reefs may be initially weakened by boring organisms such as the mollusc *Lithophaga sp.* and the sponge *Cliona sp.* Large colonies of corals such as *Porites solida* can be broken during the occasional storms whose winds are opposed to the prevailing direction.

If one considers the distribution of various growth forms of corals on a reef, an apparent paradox emerges since the delicate, ramose forms tend to occur at shallow depths where water movement is greatest while the strong, robust, massive corals seem to favour deeper water, despite the fact that they are probably better able to withstand strong wave-action. Part of the explanation for this has already been provided in our discussion of the importance of light on calcification rates, and an additional factor relates to the comparable resistance which different growth forms exert on the rate of water flow across their surfaces. In shallow water current velocities can be so high that coral polyps would be inhibited from extending and feeding unless the rate of flow was reduced. Greatest hydrodynamic resistance is provided by branching colonies which can thus reduce the water flow rates to more acceptable levels for polyp expansion. A corollary is that in areas where very weak water currents occur the growth forms with minimum hydrodynamic resistance may be favoured, since these will receive a better water exchange than those which present a higher resistance. The growth form which offers least resistance to water flow is a rounded hemispherical shape, and corals with this form do tend to dominate in areas of low water currents.

High rates of sedimentation inhibit many corals from settling or growing. Nearly all corals require a hard substrate for settlement of their planulae larvae. Many corals are able to remove moderate quantities of sediment in order to prevent clogging of their polyps and the most proficient species in this regard are also the ones which predominate an area where sedimentation rates are high.

Apart from a tendency to clog tissues or smother corals, suspended sediments also reduce light penetration and thus corals found in shallow murky waters may be similar to those found in deeper water on more open reefs, where water clarity permits much greater light penetration. This may be observed in many sheltered bays and harbours along both shore-lines of the Red Sea.

The city of Suakin was built from fossil corals and is now crumbling to coral sand.

Exposed reef top showing *Acropora variabilis* and *Stylophora pistillata.*

Structure of Corals

In this chapter we are concerned only with hard corals belonging to the Class Zoantharia and the Order Scleractinia. While some of these are solitary in growth form, most are colonial and form large reef-building colonies. The basic living unit in both cases is the coral polyp, which secretes and in turn occupies a calcareous cup or calix. Colonies are formed by the asexual budding of coral polyps and the consequent spreading of the colony. The structure of a typical coral polyp is illustrated in figure 15.

The calcareous skeleton or corallite is secreted by the basal region of the polyps. This is normally in the form of a cup in which a number of radiating vertical partitions or septa arise from the basal surface. These septa may be fused centrally to form an axial structure which is known as the columella. As the polyp continues to secrete calcium carbonate, the septa and surrounding tubular walls which form the corallite are extended until, at regular intervals, a new basal plate is laid down and the process continues once more, thus enlarging the colony size. Separate polyps are interconnected by an extension of the polyps wall (the coenenchyme) and the lower surface of this secretes the calcareous skeleton (the corallum) which develops between the individual corallites. The thickness of this coenenchyme varies in different species, so that it may be almost invisible in *Acropora* species, while it can be quite thick and easily visible in corals such as *Lobophyllia* and the brain coral *Platygyra*.

Coral species have been separated from each other and scientifically classified according to basic characteristics such as polyp size, form of budding and the structure of their skeleton, rather than by differences in growth form which, as we have discussed above, may be more related to environmental factors than to true genetic differences.

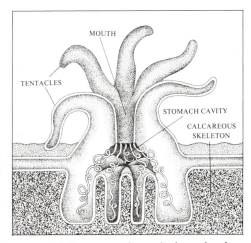

Figure 15: Structure of a typical coral polyp.

Acropora, Porites, Seriatopora and *Goniopora*

Descriptive Notes on Red Sea Corals*

Suborder: Astrocoeniina

FAMILY: ASTROCOENIIDAE
Genus: *Stylocoeniella*
Small corallites with prominent projecting septa and columella. Pillar-like structures project from surface of corallum between corallites. Polyps have a single ring of tentacles, usually twelve or less. Encrusting corals occurring sporadically from the reef top to at least 50m in depth.

Represented by: *S. armata* which has twelve or nearly twelve septa, and *S. guentheri* which possesses prominent septa and a weakly developed secondary cycle.

FAMILY: THAMNASTERIIDAE
Genus: *Psammocora*
These are encrusting, explanate, ramose or massive corals lacking a well-developed thecal wall. Septa ramify at the periphery, confluent between calices and are often petaloid. Low collines often present between calices. Columella styliform. *P. contigua* lives on the reef-top, never deeper than 5m. It is free living and its form depends upon the degree of exposure of its localines.

Represented by: *P. contigua, P. nierstraszi, P. profundacella, P. haimeana, P. explanulata.*

Opposite: coral reef with *Millepora, Stylophora, Acropora* and *Porites* sp.; above: *Stylophora pistillata.*

FAMILY: POCILLOPORIDAE
This family includes four genera in the Red Sea, i.e. *Stylophora, Seriatopora, Pocillopora* and *Madracis.* They are discussed below.

Genus: *Stylophora*
Branched corals with conspicuous corallites arranged all over the branches. Septa and columella are present. Colonies pale brown with pale branch tips or pinkish or purplish with white ends. The genus is very common in the Red Sea both in its abundance on the reefs and in the number of species which are present. Members of the genus tend to show considerable variation in growth form depending on habitat, and in the past this has given rise to some confusion regarding separation of species.

Six species are here recognised as occurring in the Red Sea: *S. pistillata, S. danae, S. subseriata, S. kuehlmanni, S. wellsi* and *S. mammillata.*

Of these, *S. pistillata* occurs in a wide range of habitats from the reef top to at least 25m; *S. wellsi* is restricted to exposed areas of the reef top; *S. danae* prefers sheltered habitats such as in harbours or among mangroves; *S. subseriata* occurs from shallow water to around 40m; *S. kuehlmanni* is a recently described thin branched species (Scheer & Pillai, 1983) which was found on reefs below 20m, and *S. mammillata* was also recently described by the same authors and was collected on Wingate reefs (25-40m).

Genus: *Seriotopora*
Slender-branched corals which are often coalescent. The coenenchyme is solid and corallites are arranged in longitudinal rows along the length of the branches. Septa and columella are rudimentary and thecal rims are extended to form hoods over the corallites. The most widespread species is *S. hystrix* which occurs in a variety of habitats from the shallow sublittoral down to around 30m depth. It has tapering narrow coalescent branches which usually have finely pointed ends. *S. caliendrum* has coalescent branches which are somewhat narrower and longer than those of *S. hystrix.* Unlike those of the latter, the branch ends are blunt. It is found in the northern Red Sea and Gulf of Aqaba. *S. octoptera* has eight to ten wing-like expansions at the branch tips. It is also found in the northern Red Sea, and Gulf of Suez.

Seriatopora hystrix

*Taxonomy of Red Sea corals has been recently reviewed in detail by Scheer & Pillai (1983). The following brief comments are based partly on their findings as well as those of Head (1980).

F. JACK JACKSON

Seriatopora hystrix

Pocillopora verrucosa

JOHN MURRAY

Genus: *Pocillopora*

These are usually branched corals with calices not arranged in rows. The surface has verrucae. There are normally twelve poorly developed septa. There are two species present in the Red Sea and they can be quite readily distinguished from one another.

P. damicornis has branches which are mainly round in cross-section, while those of *P. verrucosa* are broad and flattened. Both species occur in shallow exposed sites. On central Red Sea reefs, *P. verrucosa* is probably more abundant than *P. damicornis* and it has two colour phases: brown and purple. The brown form becomes more dominant at depths greater than 2 metres.

Genus: *Madracis*

Branching or rounded corals. Well-defined calices are polygonal or rounded. Septa in cycles of eight or ten. Columella styliform. This genus is confined to deep-water and is represented in the Red Sea by *M. interjecta* which has long thin and sometimes contorted branches. Dried fragments of the corallum are reddish. It forms quite extensive reefs at considerable depths (down to 350m) in the northern part of the Gulf of Aqaba.

FAMILY: ACROPORIDAE

There are three genera of this family in the Red Sea. They are *Acropora*, *Montipora* and *Astraeopora*. Axial corallites are only present in *Acropora*.

Genus: *Astraeopora*

Only *A. myriophthalma* is present in the Red Sea. It has an encrusting growth form and is normally found at depths greater than 14m. Corallites are distinct, separate, round in cross-section and protrude somewhat in the shape of a truncated cone. Septa are very narrow at the top and the area between the corallites has a roughened texture.

Genus: *Acropora*

Generally branching corals with the tips bearing an axial corallite. Surface coenenchyme reticulate, spinose or pseudocostate. Septa in two cycles. Columella absent.

This is one of the most common reef-building corals of all tropical seas. From the taxonomic viewpoint, it presents many difficulties since growth forms of a single species may vary considerably according to ecological conditions.

Fifteen species have been recorded from the Red Sea: *A. valenciennesi, A. pharaonis, A. nobilis, A. haimei, A. nasuta, A. corymbosa, A. hyacinthus, A. eurystoma, A. humilis, A. variabilis, A. squarrosa, A. hemprichi, A. forskali, A. granulosa* and *A. capillaris*.

However, in his recent ecological investigation of Red Sea corals, Head (1980) records only seven species: *corymbosa, cytherea, tenuis, haimei, humilis, hemprichi* and *variabilis*. Of these he notes that *cytherea* may be conspecific with *A. hyacinthus*, and *tenuis* is probably the same as *A. eurystoma*.

Shallow water *Acropora* species are often of the "stagshorn" variety while the plate-like forms tend to be more common as depth increases. Indeed, on the wide gently sloping terraces which frequently extend out from the first steeply inclined reef-face, huge *Acropora* tables may sometimes dominate the underwater scenery.

Acropora sp.

Acropora sp.

Acropora sp.

Stony corals mostly staghorn *Acropora* sp.

A. corymbosa forms dense colonies in fairly shallow exposed sites. *A. cytherea* forms large tables at medium depths. *A. haimei* occurs in very shallow water where it often covers large areas of reef. *A. humilis* tends to form knobbly clumps in the upper 8m depth range. *A. hemprichi* occurs from shallow to deep water and has both blue and yellow phases. *A. variabilis,* as its name suggests, shows marked variation in growth form with depth. *A. eurystoma* is a common species, mainly confined to the 0-8m depth range but sometimes occurring down to around 20m.

In the Gulf of Aqaba a *Acropora hemprichi/variabilis* zone extending from 13-19m has been described, but it is less than clear whether such a well-defined zone really exists in the central Red Sea. *A. hemprichi* is always more common in shallow water and *A. variabilis* (ecomorph *clavigera*) is present at depths greater than 20m. A better defined *Acropora* zone exists in shallow water on the exposed sides of many Red Sea reefs, where an association of the branching corals *A. corymbosa, A. humilis* and *A. haimei* is frequently found to cover large areas and to be mixed with *Stylophora wellsi* and *Pocillopora verrucosa*. *Millepora* is also present in this zone. This *Acropora* dominated shallow water zone has been described by Head (1980) as the Exposed Crest Zone which corresponds to Rosen's (1975) *Acropora* association.

It is beyond the scope of this book to enter into more detailed discussion of the characteristics of individual *Acropora* species.

Montipora sp.

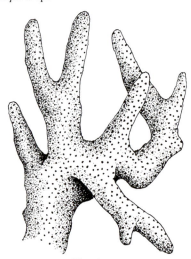

Montipora sp.

Genus: *Montipora*

Corals of this genus may form leafy, encrusting, plate-like, branching or semi-massive colonies. There is an equally wide range of colours. Corallites are about 0.5 to 1mm in diameter. Septa are in two cycles and are small or rudimentary. The skeleton is highly porous.

Fifteen species are known from the Red Sea: *M. venosa, M. spumosa, M. tuberculosa, M. monasteriata, M. verrucosa, M. meandrina, M. edwardsi, M. spongiosa, M. gracilis, M. circumvallata, M. stilosa, M. ehrenbergi, M. granulosa, M. effusa,* and *M. verrilli.*

Of these, *Montipora meandrina* is locally common in deeper water but may occur from around 5m to 30m or more. *M. tuberculosa* may also be quite common in deeper water and in sheltered bays or harbours. *M. monasteriata* and *M. verrilli* are shallow-water forms not usually found below 10m. *M. erythraea* is a locally common species which may form large overlapping colonies in around 15m depth, usually in sheltered locations. *M. venosa, M. tuberculosa* and *M. granulosa* were all found growing on reefs inside Port Sudan harbour, and the genus is one of relatively few scleractinian corals which occur in the rather turbid waters of Melita Bay on the Eritrean mainland in the Southern Red Sea.

Close-up of *Pavona decussata*.

Suborder: Fungiina

FAMILY: AGARICIIDAE

This is represented in the Red Sea by four genera: *Pavona, Leptoseris, Pachyseris* and *Gardinoseris*.

Genus: *Pavona*

Hermatypic, encrusting, columnar, foliaceous or branching corals in which the calicular wall is not differentiated, and septa are confluent between adjacent centres with beaded or serrated edges and granular sides. Septa alternate in size and are visible as fine lines running from one calix centre to another. Columella is styliform or compressed. Most colonies are pale brown.

Seven species are known from the Red Sea: *P. explanulata, P. varians, P. yabei, P. decussata, P. cactus, P. maldivensis* and *P. divaricata*.

Of these, *P. varians* occurs most frequently in sheltered environments as does *P. divaricata*. *P. cactus* forms large irregular colonies in deep water, and *P. explanulata* forms mainly encrusting (sometimes lobed) colonies at medium depths on moderately exposed reef-faces.

Genus: *Leptoseris*

The majority of *Leptoseris* corals have an encrusting base with extensions which form overlapping flattened leaves or ascending leafy scrolls. Some are entirely encrusting. In the case of flattened leafy growth forms, calices are only present on the upper surface, while in those which form upright scrolls, calices are present on both surfaces. The calices are frequently swollen and protruberant, with their boundaries indistinct and septa continuous between adjacent calice centres.

Seven species are known to occur in the Red Sea: *L. scabra, L. explanata, L. mycetoseroides, L. tenuis, L. hawaiiensis, L. fragilis* and *L. gardineri*.

In addition, a recently described closely-related species *Craterastrea levis* has been recorded from below 40m by Stephen Head, who claims that several previous Red Sea records of *L. hawaiiensis* have in fact referred to this new species. Most species seem to prefer deep water or sheltered shallow habitats. They frequently occur in caverns or overhangs where they are protected from both bright light and strong water movement.

Leptoseris gardinieri with *Goniopora* sp. in background.

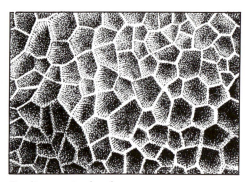

Gardineroseris planulata

Pachyseris speciosa

Pachyseris rugosa

Genus: *Gardineroseris (Agariciella)*

These are encrusting corals which tend to become massive. There is a single species in the genus and it is present in the Red Sea. *G. planulata* (previously known as *Agaricia planulata)* is an encrusting coral in which young colonies have their edges free so that they resemble a leaf, while older ones may form massive and columnar colonies. Calices are polygonal, irregular or elongate and they are closely packed with adjacent corallites, sharing walls which are narrow and prominent.

Stephen Head (1980) has recently argued that the genus *Agariciella* has priority over *Gardineroseris* and that *A. ponderosa* is the correct designation for this species. This view has been contradicted by Scheer & Pillai (1983) who claim that *G. planulata* is correct. Whichever is the case, the coral has a generally sporadic distribution from the reef top to at least 25m depth and occurs mostly in shaded habitats.

Genus: *Pachyseris*

Members of this genus form leafy, encrusting or massive colonies. The surface has concentric or irregular low collines enclosing broken or continuous lamellar columella. Calices are indistinct. The septa are closely packed together in parallel lines and their edges have extremely fine serrations.

The genus is represented by two species: *P. rugosa* and *P. speciosa*.

The former is encrusting or plate-like, frequently with vertical plates, ridges or columns. The surface of the coral rises to irregular hillocks. Collines are discontinuous and 3 to 5mm in height. *P. speciosa* generally forms leafy colonies which are lower (only rising by 1 to 2mm) and much more regular in form than those of *P. rugosa*. Colonies are generally pale to dark brown in colour.

These corals prefer sheltered conditions such as may be obtained at shallow depths on reefs inside bays or harbours and at greater depths (e.g. below 25m) on less protected reefs.

FAMILY: SIDERASTREIDAE

This family includes corals which form encrusting or rounded colonies, generally small in size. They may resemble small brain corals. Two genera are present in the Red Sea: *Siderastrea* and *Coscinaraea*.

Genus: *Siderastrea*

A single species is present, *S. savignyana*. This forms thick encrusting or massive colonies. The polygonal calices are crowded together with shared walls and their centres are depressed to give the appearance of a pitted surface.

This species is particularly tolerant of silty conditions and is widely distributed on shallow sheltered reefs as well as on the muddy bottom of lagoons, where it is one of the few species to survive healthily at depths in excess of 7m.

Genus: *Coscinaraea*

This genus also has a single species present in the Red Sea, *C. monile*. It has less regular calices than those of the previous species, but shows a number of similarities to it in the closely packed calices showing interconnecting walls and the pitted surface resulting from depressed fossae. In some cases adjacent calices merge together to give a semi-meandrine appearance.

Colonies may be encrusting or hemispherical and there is considerable variation in the species both in regard to the arrangement of calices and in the growth form of colonies. Specimens in which several columellae fuse to create a somewhat brain-like effect seem to be more common in sheltered zones where sedimentation rates are relatively high.

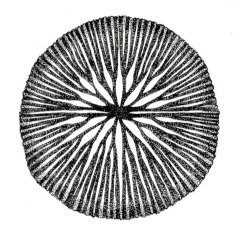

Cycloseris sp.

FAMILY: FUNGIIDAE

The family includes the solitary corals of the genera *Cycloseris*, *Fungia* and *Ctenactis* together with the colonial fungiidae *Podabacia* and *Herpolitha*.

Genus: *Cycloseris*

These have relatively small, flat discoid corolla, usually less than 5cms in diameter. The costae and septa have fine grain like teeth. Live corals nearly always have their tentacles retracted during daytime.

Six species are recognised in the Red Sea: *C. patelliformis*, *C. doederleini*, *C.cf. erosa*, *C. distorta*, *C. cyclolites* and *C. costulata*.

They seem to prefer relatively sheltered environments, such as on harbour reefs or in medium water depths on less sheltered reefs, and usually occur in sandy or rubble strewn areas close to the base of the reef slope. Juveniles are attached by a narrow stem to a hard substrate but the disc later detaches from this, leaving a small scar on its underside. Underwater, it is easy to confuse *Cycloseris* with *Fungia*. *Cycloseris* is generally smaller than *Fungia* but the primary differences are in the structure of the skeletal elements. In *Cycloseris*, the wall is imperforate whereas that of *Fungia* is perforated. Another major difference is in the edge of the costae, which are finally serrated in *Cycloseris* and markedly denticular in *Fungia*.

Fungia sp.: attached juvenile.

Fungia sp.

Genus: *Fungia*: Mushroom Corals

As I have indicated above, members of this genus are generally similar to *Cycloseris*, but they are larger and have well-defined denticulations or spines along the free edge of their costae. They may be discoidal or oval and flat or convex.

Ten species are reported from the Red Sea: *F. scutaria*, *F. moluccensis*, *F. granulosa*, *F. repanda*, *F. concinna*, *F. danai*, *F. scruposa*, *F. horrida*, *F. klunzingeri* and *F. fungites*. An additional species which is frequently listed is *Ctenactis echinata* which was previously included in the genus *Fungia*.

Of the above *Fungia* species, *F. fungites* is relatively common from the reef-top down to around ten metres depth while *F. concinna* may be common below this level.

Fungia sp. and *Ctenactis echinata* (elongate)

Ctenactis echinata

Ctenactis echinata

Genus: *Ctenactis*

As I have mentioned above, *C. echinata* was previously included in the genus *Fungia* and many recent studies record it under the latter genus. Thus there are close similarities between the two genera and *C. echinata* is the only member of this genus. It is a solitary coral with an elongate shape and an axial furrow which extends for almost the complete length of the corallum. In some cases septa of opposite sides fuse, and thus divide the fossa into one or two secondary centres. This is a common species on many Red Sea reefs and generally occurs from the reef-top to around 25m depth.

Genus: *Herpolitha*

This characteristically elongate colonial and unmistakable fungiid is represented by a single species in the Red Sea, *H. limax*. It is widespread but not generally common and is most frequently found in patch reefs (10-20m) where it lives free on the sea bed. If damaged, it has the ability to regenerate and this gives rise to distortions of the basic form. It may reach 50cms in length but is generally between 10 and 20 cms long. Its colonial nature can be confirmed by a close look at the central groove which contains a series of mouths.

Genus: *Podabacia*

These are colonial fungiids whose leafy colonies are attached to the substrate. They form encrusting plate-like, foliaceous or large bowl-shaped corolla. The underside is costate and the corallum is porous and generally brown in colour. Secondary calices are arranged around a large central calyx. Septa and costae are confluent and their margins are dentate.

P. crustacea occurs in the Red Sea, generally at depths greater than 20m. It is not a particularly common species but is widespread in its occurrence.

FAMILY: PORITIDAE

This family is represented in the Red Sea by three genera: *Porites, Goniopora* and *Alveopora*. Of these, *Porites* is the most significant in terms of its important role as a reef-builder.

Genus: *Alveopora*

There is a wide range of growth patterns from plate-like, to columnar, sub-massive and branching forms. Septa are composed of thin spines. The coral wall is extremely porous.

Five species are regarded as occurring in the Red Sea: *A. daedalea, A. verrilliana, A. ocellata, A. mortenseni, A. viridis* and *A. superficialis*.

Of these, *A. daedalea* is perhaps the most common species. It forms knob-like colonies with a diameter up to about 6cms. It tends to occur sporadically from shaded habitats in shallow water to at least 35m depth. Dead specimens are unmistakable as a result of their high porous skeleton, but living corals observed underwater can sometimes be mistaken for those of *Goniopora* or *Porites*.

Goniopora sp.

Genus: *Goniopora*

These are encrusting, massive or columniform corals which have polygonal or rounded corallites and septa arranged in three main cycles.

They often have their polyps extended during daytime and in some cases the polyp column can be as long as 2-3cms. They bear twenty-four simple tentacles which are usually spread out in a flower-like arrangement. Walls between corallites are shared and prominent. In some cases the corallites may be as small as those of *Porites* and this can give rise to confusion. Upon close examination it is possible to notice that *Porites* has less septa (about twelve) than *Goniopora* (usually twenty-four).

Seven species are present in the Red Sea: *G. stokesi, G. planulata, G. tenella, G. minor, G. savignyi, G. klunzingeri* and *G. somaliensis*.

Of these, *G. savignyi* forms large rounded colonies usually below 10m depth, and *G. klunzingeri* is probably the commonest species generally occurring on reef slopes between 20 and 30m.

Goniopora sp.

Genus: *Porites*

This genus is familiar to most people who have snorkelled or dived on Red Sea coral reefs, since it forms huge rounded light brown colonies near the reef crest of less exposed reef faces. Its surface has an almost smooth feel to it and is covered with closely packed round or polygonal calices (1.0 to 2.0mm in diameter) which remind one of a honeycomb. Septa are in two cycles.

Porites sp.

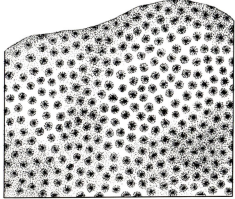

Porites sp.

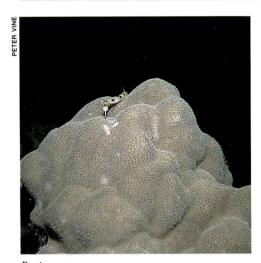

PETER VINE

Porites sp.

Ten species are known to occur in the Red Sea: *P. solida, P. lutea, P. columnaris, P. echinulata, P. punctata, P. nodifera, P. compressa, P. iwayamaensis, P. undulata* and *P. andrewsi.*

In addition, *P. lichen* has been found in deep water but Head (1980) suggests that this may be an ecomorph of *P. solida.*

P. solida is responsible for forming the huge rounded colonies which occur in shallow water on the sheltered side of many patch reefs in the Central Red Sea.

P. lutea (recorded by Vine & Head, 1977, as *P. somaliensis*) has been found colonizing the Cousteau underwater garage at Shaab Rumi reef and has a widespread distribution in the Red Sea. It may be separated from *P. solida* by virtue of the fact that the septa of the ventral triplet remain free at their ends in *P. solida,* but are united to form a trident in *P. lutea.*

Divers may confuse some species of *Montipora* with *Porites* underwater. It is usually possible to confirm *Porites* identification by the absence of any spiny tubercles or protrusions on the surface of the coral. Closer inspection will reveal other differences, especially the presence of a columella in *Porites* and the usual absence of one in *Montipora.*

On many reefs in the Red Sea, *Porites solida* is the dominant shallow water coral on the sheltered sides of reefs. In such locations it forms what may be described as the *Porites Ridge* (or the Sheltered Shallow Association — Head, 1980). On somewhat more exposed reef faces, a *"Porites zone"* occurs at around 8m or so below a shallower *Acropora* zone.

Several *Porites* species can withstand quite high turbidity and may flourish where few other corals grow. The genus is common in many harbour reefs and in more sheltered shallow water habitats such as around some of the Dhalak islands, among the Farasan bank (where *P. solida* persists into the sea-grass beds) and in the Gulf of Suez.

Suborder: Faviina

FAMILY: FAVIIDAE
Genus: *Caulastrea*

The genus is so named from the Greek "kaulos" meaning cabbage stalk and "aster" meaning star. Corallites frequently project in a form reminiscent of a cabbage stalk, and septa are arranged in a star-like fashion. Corals of this genus are usually rather low growing forms in which corallites branch and are separate from each other, so that they rise up as near parallel branches (described as a "phaceloid" growth mode). Calices are relatively large (1-2cms in diameter) and are well separated from each other at their distal extremities.

This is a genus which is fairly easy to distinguish as a result of its freely branching corallites. Although *Euphyllia* sometimes displays a similar growth pattern, *Caulastrea* can be separated by its smaller tentacles and serrated septal margins (contrasting with smooth ones in *Euphyllia*).

It is represented in the Red Sea by the relatively uncommon species *C. tumida* which has been found growing at 40m in the Gulf of Aqaba (Fara'un Island).

Genus: *Erythrastrea*

This recently named genus is closely related to *Caulastrea* but differs in its distinctive growth form, which consists of upright narrow colonies with closely packed meandering calices, which are fused throughout their length and form a sinuous arrangement of calices at the distal end. The single species so far described — *E. flabellata* — is known from the North Red Sea and Gulf of Aqaba, but future studies will most likely extend its recorded distribution to other parts of the Red Sea.

Genus: *Favia*

This is an important genus of Red Sea corals in which species may form encrusting, massive or columnar colonies. The genus is named from the Greek word "favus" which means honeycomb and the well-ordered arrangement of corallites does indeed support this description. Corallites have their own walls and are usually, but not always, quite

Favites sp.

F. JACK JACKSON

discrete from each other. Since the polyps bud by intratentacular reproduction, a distortion of the normally round calices often occurs. Septa
are alternating in width with dentate edges.

Nine species are recognised from the Red Sea: *F. stelligera, F. laxa, F.
helianthoides, F. pallida, F. amicorum, F. speciosa, F. favus, F. wisseli*, and
F. rotundata.

Of these, *F. stelligera* is a reef-top and shallow water species which is
characterised by its small low corallites. Hemispherical colonies of *F.
laxa* and hillocky growths of *F. helianthoides* usually occur below 5m and
extend down reef slopes to much greater depths in excess of 30m.
Encrusting and submassive colonies of *F. favus* and *F. speciosa* together
with globular ones of *F. pallida* occupy a broad spectrum of habitats
from sheltered reef crests to deeper locations on many types of reefs
throughout the Red Sea. *F. favus* is one of the relatively few corals to
inhabit some of the turbid waters of sheltered inlets (mersas) along the
Red Sea coastline.

Genus: *Favites*

Favites corals may form encrusting, massive, hillocky or rounded
colonies. Deep water ecomorphs are sometimes flattened and plate-
like. Corallites are fused with a raised common wall. Prominent septa
generally pass over the walls, uninterrupted between adjacent calices.
Septal margins are spiny and make the coral surface feel rough.

Seven species are present in the Red Sea: *F. peresi, F. abdita, F.
complanata, F. flexuosa, F. halicora, F. acuticollis* and *F. pentagona*.

Of these, *Favites peresi* forms hemispherical or rounded colonies
which are occasionally unattached. The inter-calice wall is acute at the
summit, thickening lower down and surrounding deeply inset fossae. It
generally occurs below 8m depth.

Like *F. peresi, F. abdita* grows in a variety of ways according to its
habitat. It may form massive encrusting or foliaceous colonies. Polygonal corallites are 7-9mm wide and 3-7mm deep. There are up to sixty
septa of which eighteen to twenty reach the columella. It may occur on
the reef top, and in fairly shallow water on sheltered reefs, and is also
found in harbour reefs.

F. complanata forms encrusting colonies with generally pentagonal
corallites (10 to 14mm wide). Usually twenty-two to twenty-five main
septa in a large calyx with the same number of secondary septa. It
occurs on reef faces from the shallows to at least 25m deep.

F. flexuosa forms submassive colonies and is widely distributed mainly
occurring in shallow water.

F. halicora demonstrates a wide degree of variation in growth forms
including encrusting, submassive and massive corolla. It lives in a broad
range of habitats and has been found from the surface to at least 20m.

F. acuticollis is not particularly common whereas *F. pentagona* is
widespread and occurs frequently in shallow reef habitats forming flat,
encrusting and sometimes hillocky colonies.

Close-up of *Favites* sp.

Below: close-up of *Favites* sp; left: *Favites* sp.

Close-up of *Platygyra daedalea*

Genus: *Goniastrea*

This usually forms rounded, convex or lobed colonies although it may sometimes have an encrusting growth-form. Closely packed corallites share interconnecting walls and calices may be polygonal and single centred (e.g. *G. retiformis*), or consist of two or three interconnected centres which create a meandering shape (e.g. *G. pectinata*). In the extreme case calices may be extended to create much more of a meandroid effect in which valleys are of variable length (e.g. *G. australensis*). The above three *Goniastrea* species are the only ones recognised as occurring in the Red Sea. *G. retiformis* tends to be a reef-top and shallow water species while both the others occur in a wide variety of habitats from near the surface to at least 30m.

Genus: *Platygyra:* Brain Coral

These generally form massive, round, brain-like colonies (up to a metre or more in diameter) in which the valleys are enclosed between thin, perforate, acute collines. Small mouths can be seen along the centre of the valleys.

Three species are recognised in the Red Sea: *P. daedalea, P. sinensis* and *P. crosslandi.*

Of these, *P. daedalea* incorporates *P. lamellina* which has often been regarded as a separate species, but in the opinion of Scheer & Pillai (1983) the two forms are merely ecomorphs of the same species. Other workers have recently argued against this view but I have here adopted the approach that *P. daedalea* and *P. lamellina* are the same species. It occurs in a wide range of habitats from the reef-top down to at least 30m. *P. sinensis* is less common but also quite widespread in its distribution. Unlike the corallites of *P. daedalea,* those of *P. sinensis* usually have a single mouth and are polygonal (about 4mm wide). *P. crosslandi* also has short valleys and single mouth corallites, but it may be distinguished by its swollen septa bearing frosted teeth which give a rough appearance to the surface of the corallum.

Platygyra daedalea

Genus: *Leptoria:* Brain Coral

This genus also includes rounded brain-like corals with sinuous valleys which are longer and narrower than those of *Platygyra*. The two genera can be more certainly separated by close investigation of the columella, which is wide and spongy in *Platygyra* but thin and lamellar in *Leptoria*.

Leptoria phrygia is the only Indo-Pacific species in the genus and it is present in the Red Sea, often on the reef top or in very shallow water, where it forms huge rounded colonies several metres in diameter.

Genus: *Oulophyllia*

This genus also forms convex or rounded meandroid colonies. Corallites are arranged in short discontinuous valleys which are wider than those of *Platygyra* (i.e. 10 to 20mm wide). Another readily observable difference between the two genera is the smaller number of septa per centimetre in *Oulophyllia* (only about six to twelve per cm compared with thirteen to sixteen in *Platygyra)*.

O. crispa is a relatively uncommon but widespread Red Sea coral which is often found on reef slopes. Colonies can reach several metres in diameter.

Genus: *Hydnophora*

In this genus, growth form is regarded as an important taxonomic characteristic. Two species are present in the Red Sea: *H. microconos* and *H. exesa*. The latter forms massive or encrusting colonies which sometimes have stout branches arising from the base. *H. microconos* is never branched but often has a thin leafy extension around the edge of the colony. The surface of the coral is unusual since the walls of corallites are raised into cone-shaped protruberances ("hydnae") which make it an unmistakable genus.

H. microconus is mainly restricted to shallow-water while *H. exesa* occurs from one or two metres deep to at least 40m.

Genus: *Diploastrea*

Colonies of this genus are usually convex or rounded with distinctive, cone-shaped and closely packed corallites. The arrangement of septa is particularly characteristic. Main septa are thickened and extend from the broad columella to the outside of the corallite where they merge with those of adjacent corallites.

There is only one species, *D. heliopora* which is relatively rare in the Red Sea. Colonies may reach a considerable size, up to several metres in diameter. Its preferred habitat is unclear since Head (1980) only found it on fringing and harbour reefs; Scheer (1964) reported on a specimen collected at Wingate reef; but Scheer & Pillai (1983) report on its presence at Shaab Anbar, and note that it was absent from collections made in the Gulf of Suez and Gulf of Aqaba. According to Wood (1983) it is a coral which is found "especially on upper reef slopes or in areas exposed to swell or currents".

Genus: *Leptastrea*

Small colonies (usually less than 25cms in diameter) may be encrusting, flat, convex or rounded, and are formed by closely packed corallites. The calices may be oval, round or polygonal and are sometimes distorted. They may show considerable size variation (e.g. from 2 to 10mm diameter within a single specimen). Numerous septa are arranged in cycles and they are not continuous with those of adjacent corallites; instead, a fine groove runs between the corallites, and this feature can be used to separate the genus from *Favites* or *Goniastrea* species with which it might otherwise be confused. Three species are present in the Red Sea: *L. bottae, L. transversa* and *L. purpurea.*

L. bottae can be readily recognised underwater by its characteristic white calices and polyps which are black in their centres. It has a wide distribution throughout the Red Sea and occurs from the reef-top to relatively deep water. *L. transversa* (which may be conspecific with *L. purpurea)* occurs in moderate depths, sometimes as encrustations on other corals (such as *Goniastrea pectinata),* while *L. purpurea* is a more common species with a wider depth distribution, occuring from the reef top to at least 30m.

Leptoria phrygia

Hydnophora exesa

Diploastrea heliopora

PETER VINE

Echinopora gemmacea with *Galaxea fascicularis.*

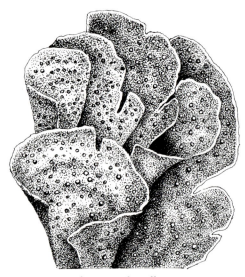

Echinopora lamellosa

Genus: *Cyphastrea*

Growth form is varied from rounded to flat, encrusting, plate-like or even branching colonies. In life they are generally light brown, sometimes tinged with pink. Corallites are regular in appearance and are small and generally well separated from each other. Cycles of septa are well-developed and quite conspicuous, with at least the first and sometimes the second cycle uniting with the columella. Two species are present in the Red Sea: *C. microphthalma* and *C. serailia.*

Of these, the former is the commonest species occurring in a wide range of habitats from the reef-flat to at least 40m. Colonies may be encrusting to submassive and hemispherical in form. *C. serailia* seems to be more restricted in its depth range, generally occurring in shallow water lagoons or on the top of reefs.

The two species may be separated from each other on the basis of their septal arrangement, the third cycle being incomplete in *C. microphthalma* but with a full complement in *C. serailia.*

Genus: *Echinopora*

The genus is named as a result of its prickly surface. The effect is mainly created by raised corallites with spiny septa, but also by the area between corallites which may be covered by spines. Colonies are usually brown or green underwater and may also be tinged with yellow or pink. Corallites are reasonably regular in their arrangement and appear as small hillocks which are raised above the surface of the coral. Two species are present in the Red Sea: *E. lamellosa* and *E. gemmacea.*

The former species has a foliaceous growth pattern, sometimes with branches whereas *E. gemmacea* is encrusting, submassive or subfoliate with corallites which project by as much as 7mm (compared to only about 1mm in *E. lamellosa).*

Both species are quite common in the Red Sea and while *E. gemmacea* displays a remarkable adaptability occurring in a wide range of habitats from the reef top to at least 40m, *E. lamellosa* is less abundant in shallow water than at deeper sites.

Genus: *Plesiastrea*

Colonies are usually massive and rounded or flattened, but encrusting forms also occur. Corallites are generally round, separate from each other with calices 2 to 3mm in diameter, slightly raised above the coral surface. There are many septa and the main ones extend across the walls and can be seen as quite conspicuous costae between the calices (a feature which distinguishes the genus from *Cyphastrea).* In some cases these costae fuse with those of adjacent calices. The edges of costae and septa are finely toothed. The genus is represented in the Red Sea by *P. versipora* which often bears a close superficial resemblance to *Favia stelligera.* It may be necessary to make a close examination of the coral to be certain of identification. Septa are more raised in *F. stelligera* and asexual reproduction is intratentacular instead of extratentacular which is more usual in *Plesiastrea.*

Wood (1983) states that this species varies its growth form according to habitat: in shallow, more exposed areas, it can form large rounded colonies, while in deeper water it often develops plate-like growths. It is not a common coral in the Red Sea and has only been recorded by two scientists: Head (1980), who found two specimens at 12m on the fringing reef near Port Sudan; and Paschke, whose collection of a single specimen in the Gulf of Aqaba is reported by Scheer & Pillai (1983).

FAMILY: TRACHYPHYLLIIDAE

Genus: *Trachyphyllia*

An unusual genus in that young corals are often solitary and older ones frequently become detached and free-living. Older specimens are small and convex shaped with large fleshy polyps. Corallites are joined so that they create a series of meandering valleys but their outer edges remain free.

There is a single species in the Red Sea: *T. geoffroyi* which is often found with young individuals attached to gastropod shells in medium depths (e.g. 15 to 50m).

FAMILY: RHIZANGIIDAE

Two genera are present in the Red Sea, *Culicia* and *Phyllangia*. They are non reef-building (ahermatypic).

Genus: *Culicia*

There is a single species, *C. rubeola,* which is a small inconspicuous coral often found attached on the undersides of other corals. In life the polyps are pink or red. They are sometimes confused with other ahermatypic corals such as *Caryophyllia* or *Dendrophyllia,* but *C. rubeola* is smaller and has corallites which are joined at their bases. In addition, their septa have dentate and not smooth edges and they do not fuse in the calices. It is probably more common than the few records of it might suggest and elsewhere it has been recorded from both shallow water and considerable depths.

Genus: *Phyllangia*

Another ahermatypic coral from deep water (115m) in the Red Sea. An unnamed species was recently collected by Prof. Fricke with his submersible "Geo".

FAMILY: OCULINIDAE

Genus: *Galaxea*

This is an important genus with a wide distribution throughout the Red Sea, where it occurs in many different habitats. *G. fascicularis* is probably the only species which is present, although *G. astreata* has also been listed and was originally described by Lamarck based upon a specimen collected in the Red Sea.

G. fascicularis shows a variety of growth forms including plate-like, massive and columnar colonies. Branching individuals are also found in some sheltered situations, and Head (1980) has suggested that it was one such specimen which confused Crossland (1939) into recording *Acrhelia* in Dongonab Bay.

The coral is green or brownish, sometimes tinged with pink. In life, transparent slender tentacles are usually extended during daytime. It has a quite distinctive skeleton formed by well separated corallites which may rise more than a centimeter above the coral surface. Blade-like septa protrude above the corallite wall.

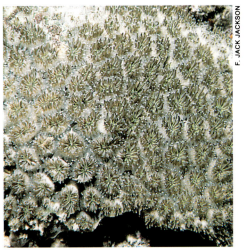

Galaxea fascicularis

Close-up of *Galaxea fascicularis*

Lobophyllia corymbosa

Lobophyllia hemprichii

Close-up of *Lobophyllia hemprichii*

FAMILY: MERULINIDAE
Genus: *Merulina*

This forms flattened thin colonies which frequently develop low hill-ocks or irregular branches. The valleys and collines are straight and spread by forking. It has a superficially meandroid appearance and calices are arranged in rows.

The Red Sea species is mentioned by Scheer (1983) as *M.cf. ampliata*, but this has been recently described by Stephen Head as a new species: *Merulina scheeri*. It is widely distributed on Red Sea reefs occurring from the shallows to at least 30m deep.

FAMILY: MUSSIDAE

Four genera from this family are represented: *Cynarina, Lobophyllia, Acanthastrea* and *Blastomussa*.

Genus: *Cynarina*

This is a solitary coral with cone or saucer shaped corallites. It is represented by *C. lacrymalis* which is cone-shaped and about 5-6cms tall. In life it is often tinged with red or green. Prominent lobed septa extend along the outside of the coral to form conspicuous spiny ridges. It favours sheltered or deeper water locations.

Genus: *Lobophyllia*

In contrast to the previously described Mussid, *Lobophyllia* forms large rounded colonies and is an important reef-builder particularly in sheltered habitats. Corallites may be single or joined to form irregular lobes or more extended meanders. The stalks may be up to 20cms or more in length and are joined at their base. Large, conspicuously serrated septa protrude by as much as 1cm above the wall and continue along the outside as low costal ridges. Two species are present in the Red Sea: *L. corymbosa* and *L. hemprichii*.

They can be separated on the basis of their corallites which are usually single-centred (occasionally two to three) in *L. corymbosa*, whereas *L. hemprichii* has corallites with long valleys. The latter species favours deeper water. It has been suggested by several workers that these are really both the same species and should be grouped under the name *L. corymbosa*.

Genus: *Acanthastrea*

This genus is represented by two species: *A. echinata* and *A. erythraea*. The former species has single mouthed corallites while those of *A. erythraea* may have one, two or three mouths. Prominent septa, with sharp teeth, are continuous between adjacent calices. There is a prominent spongy columella. *A. echinata* is not uncommon on many reefs and seems to favour moderately exposed reef-faces occurring from the shallows to at least 25m.

A. erythraea (also known as *Symphyllia erythraea*) is common along shallow, exposed, fringing reef sites along the mid-western coastline of the Red Sea but is less common on patch reefs.

Genus: *Blastomussa*

In life the skeleton of this genus is completely obscured by fleshy polyps. Rounded corallites have prominent septa whose margins bear several blunt lobes or sharp teeth. There is a solid columella.

There are three Red Sea species which are relatively easy to separate: *B. merleti* and *B. wellsi* have branching corolla while that of *B. loyae* is encrusting. The former two can be identified as a result of size (corallites 9-13mm diameter in *B. wellsi* and only 5-7mm in *B. merleti*), and the fact that a complete tertiary cycle of septa is present in *B. wellsi* but not in *B. merleti*.

B. loyae is a relatively common species, both in deep and shallow sites on fringing and patch reefs in the central Red Sea, and is also present in the Gulf of Aqaba. *B. merleti* seems to prefer fairly deep water, having been reported from 8m to 60m.

FAMILY: PECTINIIDAE

Three genera have been recorded in the Red Sea: *Mycedium, Echinophyllia* and *Oxypora*. Each of these are represented by single species, i.e. *M. elephantotus, E. aspera,* and *O. lacera*.

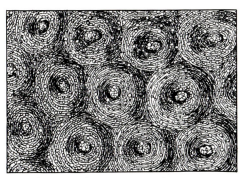

Acanthastrea echinata

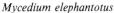

Mycedium elephantotus

J. DAVID GEORGE

Genus: *Mycedium*

Flattened, partially encrusting, colonial corals in which the growing edge rises to form the leafy extensions. Large specimens may be 2m in diameter. In life the coral is usually brown with green or pink tinges and frequently bright red or green oral discs. The coral grows by adding new calices around the edge of the corallum. Corallites are separate and protrude at an angle. Septal margins have sharp spines and numerous septa project outside the corallites into the perithecal area where they extend in narrow parallel lines towards the outer extremities of the coral.

The single species, *M. elephantotus* is frequently found in relatively deep water on steep coral slopes (e.g. 18m to 55m) and occurs throughout the Red Sea and Gulf of Aqaba.

Genus: *Echinophyllia*

This forms low encrusting or leafy colonies in which the central portion is often quite solid and firmly attached, while the outer portions are free and much thinner. Live corals are usually brown with green or pink calix centres. Slightly protruberant corallites have more vertical walls than those of the previous species. As in *Mycedium,* septa form long parallel ridges between calices but in *Echinophyllia* these are interspersed with small pits (alveoli).

E. aspera is generally found in shaded situations in shallow water and on open reef slopes down to considerable depths. (e.g. 105m at Sharm el Sheikh).

Echinophyllia aspera

Genus: *Oxypora*

Oxypora forms encrusting or flat plate-like colonies but the margins are always free and foliaceous. It is often very thin and therefore extremely easy to break when collecting it. It is usually brown in colour with the centres of calices pink, green or grey. The undersurface of the thin coral is coarsely and irregularly ribbed by spiny costae. The upper surface in life is covered by a fleshy mantle and appears somewhat warty as a result of underlying spines. Corallites are irregularly distributed with some clearly separate from each other and others joined. Ridges formed by septa run from the centre of the corallum to the periphery and continue on the underside as mentioned above. There are numerous "pores" or slits perforating the coral in young specimens, but these tend to become filled in older forms. Its general habitat seems to be similar to that of *E. aspera*.

Suborder: Caryophylliina

FAMILY: CARYOPHYLLIIDAE

Genus: *Caryophyllia*

These are solitary corals which may sometimes form clusters. The septa are in distinct cycles and costae are present. There are two Red Sea species: *C. paradoxus* and *C. sewelli*.

The former species forms fused clusters of corallites, while the latter grows as truly solitary individuals. Both species are known from 366m in the southern Red Sea.

Genus: *Trochocyathus*

This is also a deep water genus of solitary corals. Two species are represented: *T. virgatus* and *T. oahensis*. Both are quite common in deep water (e.g. 300-500m).

Genus: *Deltocyathus*

Another deep water solitary coral which is represented in the Red Sea by a single species: *D. minutus*.

Genus: *Polycyathus*

In this genus, the corallum forms small clusters as a result of external budding. They are often regarded as deep water corals but they also occur in shaded situations in shallow depths. There are two species present: *P. fuscomarginatus* and *P. conceptus*.

The former species is usually found in small clusters attached to other corals such as *Echinopora gemmacea* and may occur as shallow as 10m. *P. conceptus* tends to form branching, bushy clumps and while it is only recorded from deep water (732m and 805m) in the southern Red Sea, the author has found it growing in quite shallow water at the back of a cave in New Zealand.

Genus: *Heterocyathus*

Another deep water solitary coral usually found attached to or enclosing a gastropod shell. Represented by *H. aequicostatus* from 375m.

Genus: *Dactylotrochus*

A small, inconspicuous solitary coral often attached to *Dendrophyllia*. *D. cervicornis* has been found at 138m in the Northern Red Sea.

Genus: *Parasmilia*

Again, a deep water solitary coral. *P. fecunda* was recorded at 366m in the Southern Red Sea.

Genus: *Solenosmilia*

Known from *S. variabilis* which forms bushy colonies and was collected at 366m in the Southern Red Sea.

Genus: *Dasosmilia*

D. valida was found at 490m in the Northern Red Sea.

Genus: *Euphyllia*

E. glabrescens is found close to the lower depth limit for scuba diving. It is formed by well-separated corallites in branching colonies.

Caryophyllid: *Plerogyra sinuosa*

Plerogyra coral with *Periclimenes* shrimp.

Plerogyra sp.

Genus: *Plerogyra*

Large, well-spaced, stalked corallites with prominent septa form rounded colonies. During daytime, the skeleton is invariably obscured by balloon-like vesicles which apparently retract when the tentacles are extended at night. *P. sinuosa* is common on reef slopes and vertical underwater cliffs at moderate depths (e.g. 15m to 40m) where it seldom forms colonies larger than 50cms in diameter. Calices in this species often merge to form meanders.

Genus: *Physogyra*

This genus is uncommon in the Red Sea and was only recently reported from the area by Head (1980), who collected *P. lichtensteini* at 7-9m on a patch reef and in Port Sudan harbour. It forms rounded meandroid colonies, up to a metre in diameter, which are easy to distinguish as a result of the very prominent smooth-edged leafy septa arranged in parallel rows.

Genus: *Gyrosmilia*

This genus forms small rounded colonies in which joined corallites give rise to long sinuous valleys. A narrow groove along the top of the wall is a characteristic of the genus, which serves to separate it from *Physogyra*, which it resembles. *G. interrupta* is generally collected from the lower reef area (e.g. 40-50m).

FAMILY: FLABELLIDAE

These are solitary non-reef-building corals characteristic of well-shaded or deep water locations in the Red Sea. Represented by: *Flabellum crateriformis, Rhizotrochus typus, Javania insignis.*

All Red Sea records of them are from depths greater than 100m. They can be distinguished from other solitary corals on the basis of several characteristics, including smooth-margined septa which do not join in the calice, and the absence of costae or columella.

Dendrophylliid

Suborder: Dendrophylliina

FAMILY: DENDROPHYLLIIDAE

Genus: *Balanophyllia*

Solitary corals which sometimes bud. The only species found within range of SCUBA diving is *B. gemmifera* which occurs in deep caves along the reef edge. Other deeper water species are: *B. rediviva, B. diffusa* and *cf.B cumingii.*

Genus: *Rhizopsammia*

Small colonial corals in which colonies are formed by extratentacular budding and by stolon-like extensions from which new colonies develop. A single species, *R. wettsteini,* has recently been described by Scheer & Pillai (1983), who based their description upon specimens collected from the Gulf of Aqaba.

Genus: *Dendrophyllia*

These colonial, frequently branching and often tree-like corals are mainly found in deep water. *D. fistula, D.cf. minuscula, D. horsti, D.cf. cornigera, D. robusta* and *D. arbuscula* are all recorded from the Red Sea.

There is some debate among taxonomists as to whether the next genus should in fact be included under *Dendrophyllia,* and certain *Tubastrea* corals listed below are elsewhere considered as *Dendrophyllia* species.

Tubastrea aurea

Tubastrea micranthus

ASHOD FRANCIS

Cup coral: *Tubastrea coccinea*
Opposite: *Turbinaria mesenterina*

Genus: *Tubastrea*

Tubastrea corals may form low tufted colonies or branching tree-like ones. They are usually brightly coloured orange, red or dark green and often have bright yellow tentacles. Distinct, separate corallites have prominent septa which descend towards their centres. They generally occur where reef-building corals do not flourish, such as steep over-hanging cliffs where light is attenuated. They show a distinct preference for situations where strong currents occur, and *T. micrantha* is a particularly common species on many offshore reefs, sometimes growing in bright sunlit areas if other conditions are suitable. It forms black, branching, arborescent colonies which grow so that their branches are stretched across the direction of predominating currents.

Tubastrea aurea is another shallow water species which forms low clumps of red-orange polyps.

Other species which have been reported, but about which a certain degree of confusion exists concerning some of the records, are *T. diaphana* and *T. coccinea*. The latter has been reported by Head (1980) as occurring in caves along with *Balanophyllia gemmifera*.

Genus: *Turbinaria*

The genus is represented in the Red Sea by a single species *T. mesenterina* which shows considerable variation in growth form: often encrusting, usually with free edges, sometimes foliaceous or folded into cylindrical or funnel-shaped structures. The predominant colour underwater is yellow or brown and polyps may be brightly coloured. Rounded corallites are usually quite separate and protrude above the surface of the corallum. There are generally eighteen to twenty septa and these do not extend over the corallite wall.

T. mesenterina sometimes forms large encrusting colonies which may cover quite a considerable area of sheltered sloping reef.

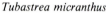

Tubastrea micranthus

F. JACK JACKSON

J. DAVID GEORGE

5. WORMS and VERMIFORM ANIMALS

PHYLUM: ANNELIDA

Class: Polychaeta

Polychaetes: Bristle worms

These belong to the Phylum Annelida which includes around 14,000 terrestrial, aquatic and parasitic species. The class Polychaeta is an almost exclusively marine group of annelids and around 8,000 species have been described worldwide. Many are cosmopolitan species with wide geographic distribution. Major habitats include burrowing in sediments, occupying the interstices of coarser deposits such as coralline fragments or living on surfaces of dead corals, algae or other organisms. Some sedentary species live in extremely close relationship with live corals, and in this regard the genus *Spirobranchus* is especially well represented in the Red Sea.

Above: Tube-worm: *Spirobranchus giganteus*
Opposite: Fan-worms: *Sabellastarte sancti-josephi (indica)*

Within the class Polychaeta there is considerable variation in the form of different species. Several features remain discernible however throughout the group. While there is no such thing as a "typical" polychaete it is possible to draw a "generalized" form which indicates the major characteristic features of the class. A diagrammatic illustration is provided in Figure 16.

A measure of the vast range of species and numbers of Polychaetes occurring in association with corals is provided by a study carried out by Dr. Frederick Grassle on a 4.7kg lump of *Pocillopora damicornis*. This contained 1,441 polychaetes belonging to 103 species.

Syllids were most abundant (921 individuals), followed by Capitellids (102 worms); Terebellids (68); Nereids (62); Lumbrinereids (56); Sabellids (44) and Phyllodocids (32). Most of the larger worms were Eunicidae. Approximately two-thirds of the macrofauna removed from the coral were polychaetes.

Although there is still a great deal to learn about Red Sea polychaetes, some progress has been made. A recent study carried out in the Northern Red Sea identified 250 polychaetes belonging to 136 genera and 26 families. In addition, Red Sea tube-worms have been investigated by the author, and numerous other workers have identified polychaetes as part of ecological studies covering a broader field of research.

The following brief notes are intended as an introduction to the range of polychaetes which are present in the Red Sea. The picture is however far from complete.

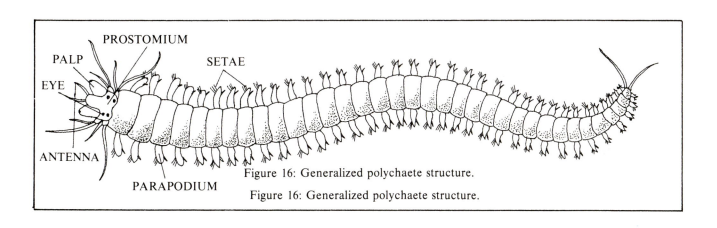

Figure 16: Generalized polychaete structure.

Figure 16: Generalized polychaete structure.

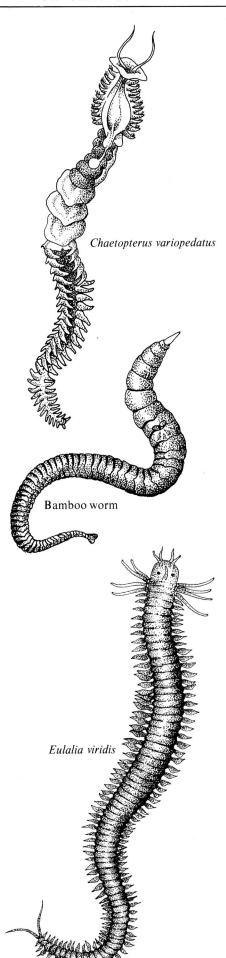

Chaetopterus variopedatus

Bamboo worm

Eulalia viridis

Order: Ctenodrilida

FAMILY: CTENODRILIDAE

These are very small sand-dwelling polychaetes which appear similar to oligochaete worms. There are no appendages on the prostomium and no tentacles or gills protrude from the body. Setae arise directly from the lateral body wall instead of from parapodial lobes. *Ctenodrilus* is a viviparous species.

Represented species in the Red Sea include: *Ctenodrilus serratus* and *Armandia intermedia*.

Order: Spionida

FAMILY: SPIONIDAE: Spionids

This is an important family which includes many species of burrowing or tube-dwelling worms which occur in most benthic environments. They have no appendages on the prostomium. The peristomium generally carries a pair (or two groups) of feeding tentacles. There are no jaws.

Represented species in the Red Sea include: *Laonice cirrata; Spiophanes koyeri; Microspio mecznikowianus; Prionospio cirrifera*.

FAMILY: POECILOCHAETIDAE

Representatives of this family include: *Poecilochaetus serpens* which occurs in shallow sandy bays.

FAMILY: CHAETOPTERIDAE: Parchment Tube-worms

These live in parchment-like tubes, open at both ends. The tubes are usually embedded in fine sediment. Enlarged paddle-like parapodia in the middle of the body draw water through a food collecting mucus net, which is secreted at the anterior of the worm. This mucus net is transferred to the mouth at regular intervals.

Represented species include: *Chaetopterus variopedatus*.

FAMILY: CIRRATULIDAE: Cirratulids

These have a pair of grooved tentacles originating from the peristomium or several pairs arising more posteriorly. They are characterized by the thread-like red branchiae (gills) which occur in some segments. They are found in sand or among loosely consolidated corraline rubble.

Represented species include: *Cirriformia semicincta; Cirratulus dasylophius; Caulleriella alata; Caulleriella caputesocis; Tharyx marioni; Dodecaceria joubini; Audouinia (Cirratulus) gracilis*.

Order: Capitellida

FAMILY: CAPITELLIDAE: Capitellids

These are thread-like worms which are coloured red anteriorly. Their prostomium lacks appendages. Thoracic and abdominal setae differ and the first one or two abdominal segments lack setae altogether. The lower portion of each parapodium forms a pad. They have an eversible thin-walled proboscis. They may be found among dead coral rubble and surrounding sediments.

Represented species include: *Dasybranchus caducus; Capitella capitata*.

FAMILY: MALDANIDAE: Bamboo-worms

These have elongated segments which resemble the pattern of a bamboo stalk. They are closely related to capitellid worms (belonging to the same order, *Capitellida*) and share most of their general characteristics. They are usually found in relatively deep water (more than 40m), in the sediment.

Represented species include: *Clymene (Praxillella) gracilis; Petaloproctus terricola; Clymene watsoni; Clymene africana; Clymene affinis; Macroclymene monilis; Clymene lombricoides*.

Order: Opheliida

FAMILY: OPHELIIDAE: Opheliids
These usually have short, grub-like bodies. Their prostomium lacks appendages and their poorly developed parapodia carry simple tapering setae on all segments except the first. They have an unarmed eversible proboscis. Opheliids are burrowing worms which are found in coral sand and among coral rubble.

Represented species in the Red Sea include: *Polyphthalmus pictus; Armandia longicaudata.*

FAMILY: SCALIBREGMIDAE
These have a T-shaped prostomium which lacks appendages. The body is frequently anteriorly inflated and skin may appear wrinkled. Anterior body segments may carry branching gills. They are found in subtidal muddy sand and in sand under coral rubble.

Order: Phyllodocida

FAMILY: PHYLLODOCIDAE: Paddle-worms
These are elongate worms with prominent paddle-like parapodia. They have no jaws but an eversible proboscis and are generally scavengers on dead or dying marine life.

Represented species include: *Phyllodoce madeirensis; Phyllodoce quadraticeps; Phyllodoce gracilis; Eulalia viridis; Eulalia tenax.*

Lepidonotus carinulatus

FAMILY: APHRODITIDAE: Sea Mice
The common name is derived from numerous hair-like bristles which cover the upper surface of these worms and fringe both sides of the body. They usually occur in soft sediments in fairly shallow water.

Other scale-worm families in the Red Sea include: *Polynoidae; Polyodontidae; Pholoididae; Eulepethidae* and *Sigalionidae.*

Represented species include: *Pholoe minuta; Eupanthalis kinbergi; Pareulepis wyvillei; Sigalion mathildae; Polyodontes maxillosus; Lepidonotus carinulatus; Harmothoe gilchristi; Hololepidella nigropunctata; Lepidonotus cristatus; Lepidonotus glaucus.*

Coral tube-worm: *Spirobranchus giganteus*

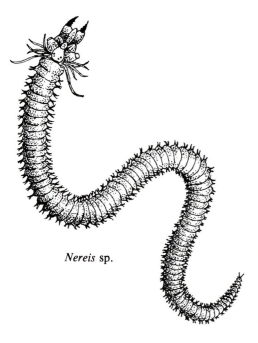

Nereis sp.

FAMILY: CHRYSOPETALIDAE
These belong to the sub-order *Aphrodontiformia* although they lack the prominent scales of their related forms. Represented species include: *Paleanotus debilis; Paleanotus chrysolepis; Chrysopetalum ehlersi.*

FAMILY: HESIONIDAE: Hesionids
These are frequently found among dead coral rubble in shallow water. They are relatively short, flattened worms which feed on a variety of small animals which are enveloped in their eversible pharynx. They have well-developed eyes.

Represented species include: *Leocrates claparedii* which occurs on dead coral and *Hesione pantherina* found in sandy bays.

FAMILY: NEREIDAE: Nereids
This family includes the ragworms which are familiar to many fishermen. They have long many-segmented bodies and an eversible pharynx with a strong pair of jaws. The cosmopolitan species: *Nereis pelagica* which occurs in the Red Sea may grow to over 10cms in length. *Perinereis nuntia* and *Naianereis quadraticeps* are found burrowing in sand in sheltered infratidal areas such as in the Gulf of Suez.

Represented species include: *Nereis mirabilis; Nereis trifasciata; Nereis zonata persica; N. couteri; N. pachychaeta; N. unifasciata. N. erythraeensis; N. costae; Leptonereis glauca; Platynereis pulchella; Perinereis nuntia; Perinereis nigropunctata; Nereis jacksoni; Leonnates jousseaumei.*

Fire-worm: *Hermodice carunculatus.*

FAMILY: PILARGIIDAE
This family is in the suborder Nereidiformia and is thus related to Syllid and Nereid worms which are of major importance in terms of their abundance in certain habitats.
Represented species: *Sigambra parva.*

FAMILY: SYLLIDAE: Syllids
These are well represented in most habitats within the Red Sea. They are small, usually threadlike worms with an eversible proboscis and they usually have a prominent feeler associated with the upper part of each parapodium. They possess three antennae and a pair of palps on the prostomium and four tentacles on the peristomium.
Represented species include: *Typosyllis armillaris; Branchiosyllis uncinigera; Opisthosyllis papillosa; O. laevis; O. longicirrata; Odontosyllis gibba; Parasphaerosyllis indica; Grubea limbata; Autolytus prolifer; Syllides fulva; Haplosyllis bisetosa; Sphaerosyllis capensis; Trypanosyllis zebra; Grubea clavata; Grubea tenuicirrata; Haplosyllis spongicola; Syllis gracilis; Syllis variegata; Syllis krohnii.*

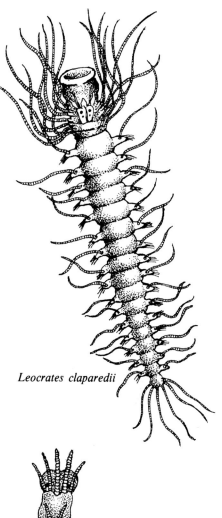

Leocrates claparedii

FAMILY: GLYCERIDAE: Glycerids
Glycerid polychaetes have two pairs of antennae on a conical prostomium. They lack feelers but have a long eversible pharynx which usually bears four black jaws. They frequently occur among coral rubble. Some species are quite large.
Represented species include: *Glycera tesselata; Glycera cirrata; Glycera africana; Goniada longicirrata.*

FAMILY: NEPHTYIDAE: Nephtids
Represented species include: *Nephtys sphaerocirrata; Nephtys malmgreni; Nephtys dibranchus; Nephtys inermis.*

FAMILY: TOMOPTERIDAE
A representative of this family found in the Red Sea is *Tomopteris (Johnstonella) dunkeri.*

Order: Amphinomida

FAMILY: AMPHINOMIDAE: Fire-worms
The common name results from sharp setae which project in tufts from the parapodia, and which tend to penetrate the skin and break off inside the flesh of those who handle them. They can cause quite intense local pain and the wounds sometimes become infected. They are quite commonly found beneath boulders and coral heads. Of particular interest is *Hermodice carunculata* which feeds at night on live corals, mainly *Porites.*
Represented species include: *Eurythoë complanata; Hermodice carunculata; Chloeia fusca; Notopygos megalops; Euphrosyne foliosa.*

Order: Eunicida: Eunicids

These include a wide range of carnivorous species, some of which are as long as 1m in length. They have several jaws associated with an eversible proboscis. They occur in sand and among coral rubble.
The unusual deep water species in the family Onuphidae, *Hyalinoecia tubicola,* lives in a tube which is similar to the quill of a feather. This lies horizontally on the seabed and it feeds by protruding from the front of the tube and scavenging among the sediment. When disturbed it withdraws and both ends of the tube are then closed by valves. *Eunice torquata* is commonly found buried in sand close to lowtide level.
Represented species include: *Eunice australis; E. aphroditois* var. *punctata; E. marenzelleri; E. torquata; Nematonereis unicornis; Oenone fulgidae; Dorvillea angolana; E. siciliensis; Hyalinoecia tubicola; Lumbriconereis coccinea; L. gracilis; Protodorivillea kefersteini; Eunice indica; Eunice antennata; Lysidice collaris; Onuphis furcatoseta.*

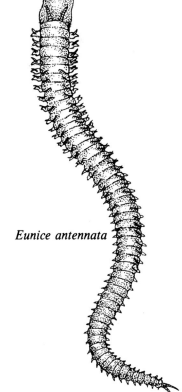

Eunice antennata

Order: Flabelligerida: Flabelligerids

In this order, the prostomium carries lateral palps and filamentous branchiae arise from the peristomium. The entire head region is retractible into the first three segments of the thorax. They secrete mucus from their bodies and thus ensheath themselves in sand grains. They have green blood. They occur among coral rubble.

Order: Terebellida: Terebellids

These tube dwellers also have long contratile feeding tentacles which may be withdrawn into the mouth in Ampharetidae but not in Terebellidae. The anterior segments often carry branched gills. Their tubes are formed by sand, shells and other fragments adhering together by mucus secreted by the worms. They may be found among coral rubble, attached to coralline fragments, and buried in surrounding sediments where their feeding tentacles extend over the substrate in search of food particles, which are trapped by mucus and passed along a ciliated groove to the mouth.

Represented species include: *Eupolymnia nebulosa; Streblosoma persica; Streblosoma cespitosa; Nicolea venustula; Loimia medusa; Terebellides stroemi; Polycirrus plumosus; Reterebella queenslandia; Terebella ehrenbergi; Terebella lapidaria; Pista typha.*

FAMILY: AMPHARETIDAE: Ampharetids

This is a family of tube dwelling polychaetes which have their body divided into two regions. The prostomium lacks appendages and the peristomium carries feeding tentacles which can be withdrawn into the mouth.

Order: Oweniida

FAMILY: OWENIIDAE: Oweniids

Members of this family live in membranous tubes which they secrete and to which sand grains and coralline fragments adhere. Their presence in the sediment or among coral rubble is often signalled by their protruding soft tubes which may project a few centimetres above the substrate surface. A frequently recorded species is *Oweenia fusiformis.*

Sabellastarte sanctijosephi = (indica)

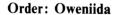

Sabellastarte sanctijosephi = (indica)

Sabellastarte sanctijosephi = (indica)

Undersides of platy corals provide settlement surfaces for a variety of invertebrates including serpulid tube-worms.

Order: Sabellida: Fan-worms

FAMILY: SABELLIDAE

These unmistakable worms secrete non-calcareous-membranous tubes in which they live and from which their feathery tentacles protrude in a fan-shaped net when the worms are feeding on their diet of small planktonic organisms. When disturbed they withdraw into their tubes. Their tentacles are frequently reddish or banded with red. They may be found attached to dead coral or partially buried in the sediment with a section of their tube protruding away from the substrate.

Represented species include: *Chone collaris; Dasychone conspera; D. cingulata; D. lucullana; Oriopsis armandi; Fabricia filamentosa; Branchiomma sp.; Sabellastarte sanctijosephi (= indica); Bispira sp.; Branchio mushaensis.*

FAMILY: SERPULIDAE: Calcareous Tube-worms

These may be divided into two main groups, i.e. those which have coiled tubes, the Spirorbidae, and those whose tubes are generally uncoiled, the Serpulidae. Their white calcareous tubes are frequently patterned with longitudinal ridges, spines or by transverse thickening. One of the major characteristics used in describing, and consequently separating, different species is the form of the operculum, which is a modified tentacle used for closing off the entrance of the tube when the worm withdraws into it. In the brief notes which follow the most frequently encountered species are mentioned and their opercula illustrated.

Serpula concharum has three longitudinal ridges on a chalky white tube. Soft, bell-shaped operculum. Occurs in shaded situations on dead coral and other substrata such as ship wrecks from 5m to at least 50m depth.

Serpula lobiancoi is unusual for this family in that the tube is generally tightly coiled — either clockwise or anticlockwise. It has a pinkish hue. The operculum is lightly chitinised. Occurs in relatively deep water (e.g. 80m) on dead coral.

Serpula concharum

Serpula lobiancoi

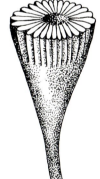

Serpula vermicularis

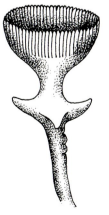

Crucigera tricornis

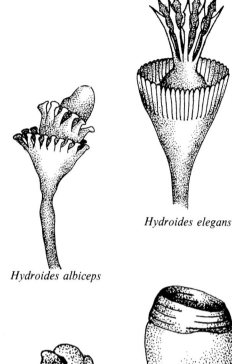

Serpula vermicularis has a pink tube, unevenly thickened with five irregular longitudinal ridges. Operculum is in shape of a fluted funnel. Found on undersides of platy corals from 10-35m.

Crucigera tricornis has a distinctive tube, unevenly thickened with five longitudinal ridges. Operculum concave cup with three prominent processes at base. Found on dead coral at 50m.

Hydroides albiceps: Tube warty and porcellanous. Operculum funnel shaped with a central crown of chitinous spines with spatulate ends and one large bulbous lobe dominating. Found under coral slabs from 30 to 50m.

Hydroides elegans: Tube thick, chalky white and smooth, may have two or three longitudinal ridges. Operculum is an inverted cone with chitinous spines and a central crown of short spines. Under coral slabs in moderate depths (e.g. 20 to 40m).

Hydroides heterocera: Tube thick, up to 3cms long. Operculum has central crown in which one spine is enlarged and recurved. Found in shallow water on coralline material.

Hydroides elegans

Hydroides minax: Tube chalky white and round in cross-section. Operculum is somewhat varied in form but consists of an inverted cone with a large recurved spine bearing three prongs at its end and other spines lower down.

Hydroides albiceps

Hydroides perezi: Tube small and flattened, chalky white with faint and irregular transverse and longitudinal ridges. Operculum as illustrated. Found under coral slabs from 3-30m.

Vermiliopsis agglutinata has a small narrow tube with 3 longitudinal ridges. Operculum a dark brown cap on a translucent ampulla. Found under corals and on wrecks from 10-30m.

Vermiliopsis infundibulum has a thick white, rough surfaced tube consisting of interlocking sections. Operculum opaque membranous bladder surmounted by a chitinous cone. Found attached to dead corals or shells (20-80m).

Vermiliopsis agglutinata

Vermiliopsis pygidialis: Tube chalky white with 3 longitudinal ridges. Operculum as illustrated. Found under coral 4m to 75m.

Semivermilia pomatostegoides: Tube sometimes coiled and has a median longitudinal ridge. Operculum is a stack of smooth-edged, round plates. Occurs under corals 30-80m.

Hydroides perezi

Apomatus sp. has a coiled tube. Operculum membranous sub-spherical sac. Found on coral rock (30m).

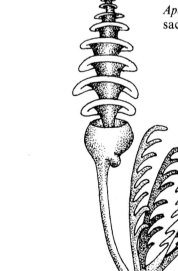

Apomatus sp.

Semivermilia pomatostegoides

Vermiliopsis pygidialis

Hydroides minax.

Spirobranchus giganteus corniculatus

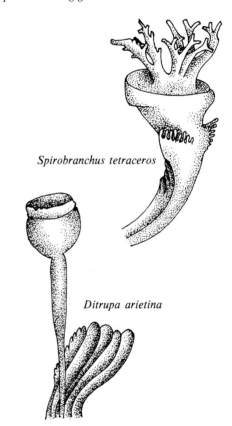

Spirobranchus tetraceros

Ditrupa arietina

Tube-worms: *Filograna* sp.

Spirobranchus giganteus corniculatus is a large tube-worm which is often encountered by divers, since its tube is embedded in live corals and its brightly coloured tentacles protrude in two fan-shaped spiral whorls from their embedded tubes, and form a prominent sight underwater. Their tubes usually have a single sharp spine above the mouth. This presumably discourages predators. Operculum has deer antler-like spines arising from the opercular plate. Found in association with live corals such as *Porites* and *Acropora,* and with the hydrocoral *Millepora.*

Spirobranchus latiscapus is a deep water species whose tube has a median dorsal ridge formed by prolonged, curved spines. Operculum has a number of stacked plates and there are two long "wings" where the stalk joins the operculum. Found under coral in deep water (e.g. 80m).

Spirobranchus tetraceros: Tube light pink and with median longitudinal ridge. Not normally embedded in coral. Operculum variable and as illustrated. Found on shipwreckage and corals *(Stylophora* and *Porites)* and on valves of pearl shells *(Pinctada margaritifera)* in shallow water.

Ditrupa arietina: Tube is not attached to the substrate and is round in cross-section with both ends open. Operculum as illustrated. Found with coral fragments in deep water (75-80m).

Filograna implexa has narrow white intertwining tubes with occasional growth rings. There is no operculum. Found in encrusting sponges, ascidians and other organisms in sheltered locations.

Filogranaella elatensis is a gregarious tube worm whose tubes aggregate to such an extent that they may form small "reefs". No operculum. Body symmetrical; twelve or more thoracic segments and up to sixty abdominal ones. Differs from *Filograna implexa* in the form of its collar setae which are limbate (as opposed to bayonet setae which occur in *F. implexa).*

Protula tubularia has a white round tube with growth rings, often coiled at base with a terminal ascending portion. Branchiae form prominent red banded fans. There is no operculum. Found in calm water, e.g. 10m in Suakin harbour and 60m on the fringing reef.

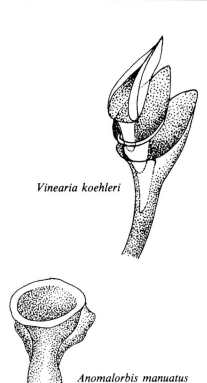

Vinearia koehleri

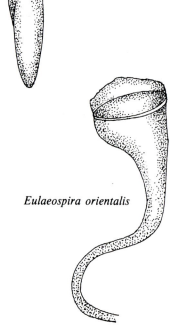

Anomalorbis manuatus

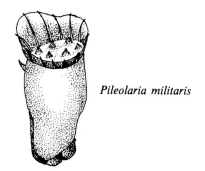

Eulaeospira orientalis

F. JACK JACKSON

Spirobranchus giganteus

Pileolaria militaris

FAMILY: SPIRORBIDAE

These are coiled clockwise or anticlockwise and are so small that they are usually overlooked. There are around ten species which are known to occur in the Red Sea: *Eulaeospira orientalis; Vinearia koehleri; Pileolaria militaris; P. pseudoclavus; Janua pagenstecheri; Janua brasiliensis; J. steueri; Neodexiospira foraminosa; N. preacuta; Anomalorbis manuatus*. Of these, special mention can be made of several species.

Eulaeospira orientalis is a winter breeding species, clockwise coiled and unlike most other tropical Spirorbids it incubates its embryos in its tube. It is very common on dead coral from 3-30m and optimum settlement occurs at around 10m.

The species *Vinearia koehleri* and *Pileolaria militaris* both have strong thick tubes which can withstand abrasion and are less likely to be scraped off rock surfaces by reef grazers than most other Red Sea Spirorbids. They both occur from intertidal or extremely shallow regions, on the reef-flat or reef crest, to about 30m depth. They attach to dead corals, coralline algae or to almost any other hard substrata.

The related species *Pileolaria pseudoclavus* favours sheltered regions and has a terminal ascending tube mouth which keeps its feeding tentacles clear of sediment or from being blocked by encrusting organisms.

Among the *Janua* species, *Janua steueri* is particularly abundant and its tube may coil in a clockwise or anticlockwise direction.

Finally, the recently described form *Anomalorbis manuatus* (Vine, 1972) is a remarkable species with characteristics which suggest an intermediate position between Serpulidae and Spirorbidae. It has the equivalent of four and half thoracic segments and is a tube incubator. To date it has only been found at 30m on pieces of the famous shipwreck Umbria which lies off Port Sudan.

OTHER WORM-LIKE ANIMALS

In addition to true Annelida there are quite a number of worm-like creatures which are found in the Red Sea.

PHYLUM: SIPUNCULA: Peanut-worms

These are unsegmented worms with quite tough, muscular bodies comprising a cylindrical trunk-region and a narrow anterior retractable introvert. When this is everted it can be observed to consist of a mouth surrounded by a ring of short, grooved tentacles. In many species the body surface is covered with small, chitinised papillae. They live in crevices or burrow in sand or mud.

Red Sea species include: *Sipunculus robustus; Golfingia vulgaris; Phascolosoma nigrescens; P. scolops; P. pacificum; Aspidosiphon elegans; A. klunzingeri; Phycosoma meteori;* and *Phycosoma ruppelli.*

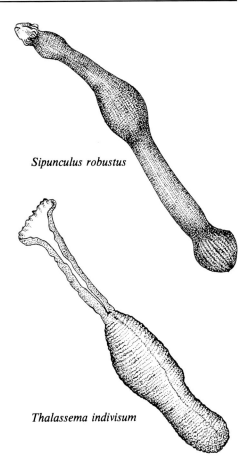

Sipunculus robustus

PHYLUM: ECHIURA: Tongue-worms

These are sausage-shaped unsegmented worms with a grooved tongue-like proboscis and a pair of chitinous hooks embedded in the ventral body wall. They burrow in mud and are detritus feeders. In the Red Sea, *Ochetostoma erythrogrammon* and *Thalassema indivisum* are both quite common.

Ochetostoma erythrogrammon is around 15cms long, reddish brown in colour with a white gelatinous mass surrounding the posterior end of the trunk. It can be quite abundant in shallow water where boulders overlie pockets of sand.

Thalassema indivisum

PHYLUM: NEMERTEA: Ribbon-worms

These are slender unsegmented "worms" which are typically a few inches long but can extend their bodies to much greater lengths. Their body surface is ciliated and secretes abundant mucus so that when they move across the surface of corals or along the sea bed a neat trail of translucent mucus marks their path. They are carnivorous, generally feeding on polychaetes which they capture by ejecting a long tubular sticky proboscis which emanates from a pore on the tip of the head.

Relatively little is known about their distribution or ecology in the Red Sea and few species have been recorded from the region. Representatives include: *Baseodiscus unistriatus; B. curtus; B. hemprichii; Nemertopsis peronea; Diplopleura obockiana.*

A typical environment for several of these species is under coral boulders in shallow water, often on the reef flat.

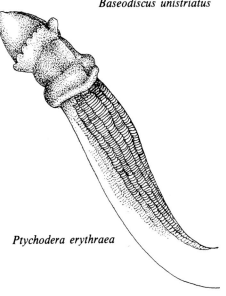

Baseodiscus unistriatus

PHYLUM: HEMICHORDATA: Acorn-worms

These are more closely related to vertebrates such as fish than to invertebrate worms but their body form is frequently superficially worm-like. They often burrow in sand and may be responsible for creating large mounds on the sandy bottoms of lagoons or sheltered areas. Their body can be divided into a proboscis, a collar and a long trunk.

Acorn worms feed on detritus and are seldom seen by divers since they make deep burrows which are quite difficult to excavate. The main species occurring in the Red Sea are: *Ptychodera erythraea* (see drawing) and *P. flava.*

The latter species may form quite dense communities in shallow soft sediments, frequently in areas where the sea-grass *Cymodocea ciliata* is present. It burrows to about 40cms deep. A *Ptychodera flava / Radianthus koseirensis:* burrowing sea-anemone community was described by Fishelson (1971) who sought to characterize various shallow water benthic communities in the Red Sea.

Ptychodera erythraea

PETER VINE

6. CRUSTACEANS

PHYLUM: CRUSTACEA

A bewildering array of species exhibiting a wide range of forms are included in the phylum Crustacea (which was previously considered as a single class of Arthropoda). All of them are characterized by segmented bodies and a chitinous covering or exoskeleton which necessitates a series of moults to permit growth. Crustaceans have numerous paired appendages and bodies divided into a head, thorax and abdomen. They have gills for respiration and there are three pairs of appendages around the mouth, i.e. the mandibles and first and second pairs of maxillae.

These may be followed by paired maxillipeds. Some of the eight thoracic segments may be fused so that the dorsal skeleton forms a single carapace. Abdomen usually has six segments which bear swimming appendages. Table V indicates the major groups which may be encountered.

Opposite: Painted Crayfish: *Panulirus versicolor;* above: Hermit crabs: *Coenobita scaevola.*

Table V					
ABBREVIATED CLASSIFICATION OF CRUSTACEANS (Covering only those forms discussed in this account)					
Phylum	**Class**	**Superorder**	**Order**	**Suborder**	**Infraorder**
Crustacea	Branchiopoda		Cladocera		
	Ostracoda				
	Copepoda				
	Cirripedia				
	Malacostraca				
		Hoplocarida	Stomatopoda		
		Peracarida	Mysidacea		
			Cumacea		
			Tanaidacea		
			Isopoda		
			Amphipoda		
		Eucarida	Euphausiacea		
			Decapoda	Natantia	Penaeidea
					Caridea
					Stenopodidea
				Reptantia	Palinura
					Anomura
					Branchyura

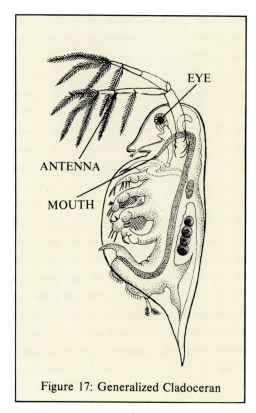

Figure 17: Generalized Cladoceran

Class: Branchiopoda

Order: Cladocera: Water Fleas

These are small planktonic crustacea which have a distinct head with a prominent sessile eye and a pair of antennae which are used for swimming. The carapace is laterally compressed. Although most Cladocerans are fresh-water organisms, several marine species are recorded from the Red Sea, where they may form an important food source for the numerous planktivorous animals associated with coral-reefs. Figure 17 shows a generalized Cladoceran.

Evadne tergestina is abundant in summer plankton. *Penilia avirostris* has been reported as common in the Gulf of Suez and elsewhere in the Red Sea. *Poden schmackeri* is also recorded. These are all widely distributed in the Indian Ocean.

Class: Ostracoda

Like the copepods and cladocerans, ostracods are also small crustaceans and marine species may be planktonic or bottom living. They range in length from 1mm to about 2cms and are characterized by a bivalved carapace which encloses the body (see figure 18).

At least twelve species are present in Red Sea plankton: *Cypridina dorsocurvata; Halocypris atlantica; Pyrocypris amphiacantha; P. chierchiae; P. rivilli; P. sinuosa; Philomedes gibbosa; P. polae; Philomedes sp.; Asterope arabica; A. mariae; Euconchoecia chierchiae.*

Benthic ostracods include: *Cycloleberis brevis; C. lobiancoi; Cylindrolebris grimaldi; Cypridina inermis; C. serrata; Cypridinotes galatheae; Euphilomedes corrugata; E. ferox; Philomedes polae; Rutiderma normanni; Synasterope empontseni.*

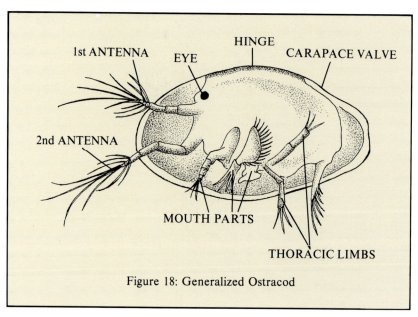

Figure 18: Generalized Ostracod

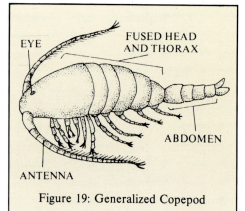

Figure 19: Generalized Copepod

Class: Copepoda

These are small parasitic or free-living crustaceans which are mainly planktonic. Parasitic forms include "fish-lice" which are often found on fish gills or in association with other marine invertebrates. They are extremely abundant and are important participants in the food webs affecting coral reefs. A generalized copepod is illustrated in figure 19.

The copepods *Anthessius amicalis* and *A. alatus* are found in the pallial cavities of *Tridacna* clam shells in the Red Sea.

At least 158 species of planktonic copepods are recorded from the Red Sea and if this seems a large number it is worth noting that there are at least 270 species of Indian Ocean copepods, many of which are prevented from occupying the Red Sea as a result of its temperature and salinity characteristics. Several species are known only from the Red Sea.

Planktonic populations of copepods vary at different times of year, while some species are present throughout the year, others are introduced during winter months when north-east monsoons cause an inflow from the Indian Ocean and some species migrate from deep water towards the surface.

In view of the large number of species which are present and the specialised nature of this subject, I have restricted the list of species to those which are distributed throughout the Red Sea basin in all seasons. For a more comprehensive review, the reader should consult Halim's review of Red Sea plankton (Halim, 1969).

Table VI
LIST OF RED SEA PLANKTONIC COPEPODS DISTRIBUTED THROUGHOUT THE RED SEA AT ALL TIMES OF THE YEAR*

Acartia centrura	*Euchaeta concinna*
A. erythrea, A. negligens	*E. marina*
Acrocalans gibber, A. gracilis	*Euterpina acutifrons*
Calanopia elliptica, C. minor	*Labidocera acuta, L. minuta*
Calocalanus pavo	*Microsetella atlantica, M. rosea*
Candacia bispinosa, C. bradyi	*Oithona plumifera, O. similis*
C. catula, C. curta	*Oncea conifera, O. media*
Canthocalanus pauper	*O. mediteranea, O. venusta*
Centropages elongatus	*Paracalanus aculeatus, P. parvus*
C. furcatus, C. gracilis	*Pleuromamma abdominalis, P. robustus*
C. orsini, C. violaceus	*Pontellopsis krameri*
Clausocalanus arcuicornis	*Rhincalanus nasutus*
C. furcatus	*Sapphirina nigromaculata*
Clytemnestra scutellata	*S. ovatolanceolata*
Copilia mirabilis	*Setella gracilis*
Corycaeus danae	*Scolecithrix chelipes*
C. gibbulus, C. venustus	*Temora stylifera, T. discaudata*
Eucalanus subcrassus	*Undinula vulgaris*

*After Halim, 1969.

Tetraclita squamosa barnacles on intertidal rock.

Chthamalid barnacles

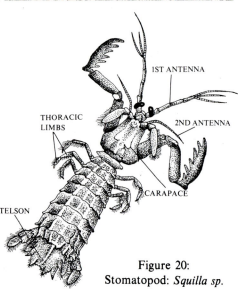

Figure 20:
Stomatopod: *Squilla sp.*

Class: Cirripedia: Barnacles

Intertidal barnacles are not as common in the Red Sea as they are in many other marine locations. This is partially due to the restricted diurnal tidal range which occurs for the greater part of the Red Sea. In those areas where a significant daily tidal fluctuation does occur the barnacles *Tetrachthamulus oblitteratus* and the giant barnacle *Tetraclita squamosa rufotincta* may form quite distinct zones with the smaller species, *(T. oblitteratus)* occurring close to the top of the shore and *T. squamosa rufotincta* extending over the major area of the barnacle zone. *Tetraclita* is more predominant on exposed shores where relatively strong turbulence occurs while *T. oblitteratus* is more tolerant to heat and dessication than *Tetraclita* and can therefore extend higher up the shore.

Other barnacles include *Chthamalus stellatus; C. depressus; Lithotrya valentiana; Ibla cumingi; Balanus amphitrite* and the Goose barnacle *Lepas anatifera.*

Class: Malacostraca

Superorder: Hoplocarida

Order: Stomatopoda: Mantis Shrimps

These large crustaceans have stalked eyes but only a small carapace which leaves the last four thoracic segments uncovered. The second leg is greatly enlarged to form a mantis-like raptorial claw (figure 20). They live in burrows in sand or mud and feed at night.

At least twenty forms are present in the Red Sea: *Squilla carinata; S. latreillei; S. gonypetes; S. massavensis; S. harpax; Pseudosquilla ciliata; P. megalophthalma; Lysiosquilla maculata; L. multifasciata; Coronida trachura; Gonodactylus chiragra; G. glabrus; G. demani demani; G. demani spinosus; G. smithii; G. falcatus; G. brevisquamatus; G. pulchellus; G. spinosissimus; G. lanchesteri; Acanthosquilla multifasciata; Protosquilla lenzi; Pullosquilla thomassini.*

Superorder: Pericarida

Order: Mysidacea: Opossum Shrimps

These small shrimp-like crustacea have stalked eyes and a body divided into head, thorax and abdomen. A carapace is fused to four thoracic segments (figure 21). They may be abundant in coastal waters where they live among algae or on muddy bottoms, rising into the water column at night. They are detrital feeders or scavengers.

Represented species include: *Siriella brevicauda; Kainomatomysis foxi; K. schieckei* and *Gastrosaccus sp.*

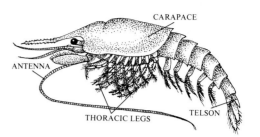

Figure 21: Opossum Shrimp

Order: Cumacea

These have an ovoid body and slender, tail-like abdomen (figure 22). They are small crustaceans, generally less than 5mm and live in sandy mud where they feed on organic detritus. At night they rise in the water and are thus caught in plankton hauls.

Represented species include: *Nannastacus gurneyi; N. spinosus; Schizotrema pori; Cumella limicoloides; C. forficuloides; Pseudocuma longirostris; Bodotria alata; Camylaspis akabensis; Cyclaspis maris rubri.*

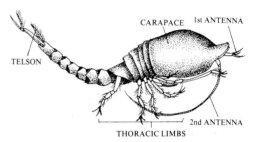

Figure 22: Generalized Cumacean

Order: Tanaidacea

These are generally of microscopic dimensions and are not well studied in the Red Sea. They are characterized by having their first walking legs modified as chelae, and a carapace covering head and first two thoracic segments (figure 23). Shallow water species are rarely longer than 5mm. They are found among hydroids, sponges and in the empty tubes of Serpulids, as well as in many other comparable niches. Marine biology students are likely to encounter them when studying collected samples under binocular microscopes.

Representative genera in the Red Sea include: *Apseudes; Leptochelia; Pagurapseudidae.*

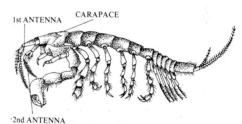

Figure 23: Tanaidacean: *Apseudes* sp.

Order: Isopoda

These have dorso-ventrally flattened bodies and lack a carapace. The anterior region has seven pairs of similar appendages, hence the name "isopod" (similar feet). Isopods include terrestrial and fresh water forms but most are marine. A generalized isopod is shown in figure 24.

In the Red Sea they are present in a wide range of habitats including sand, coral rubble, weed, and as parasites on fish and on other crustaceans. They are mainly detrital feeders. There are far too many species present to permit a complete list, and even if space was not a limiting factor, there has been relatively little research carried out on Red Sea Isopods. Species discussed below are the most significant ones in collections made on the shore line of the Gulf of Aqaba, Gulf of Suez and parts of both Red Sea coastlines (See D. Jones, 1974).

Eurydice arabica is a Red Sea endemic species which is found in the mid tidal zone of sandy beaches.

Eurydice inermis is a sub-tidal species which lives in offshore sand and gravel, and migrates towards the surface at night-time from depths as great as 50m or more. It is usually collected as a result of being attracted to light.

Cirolana rotunda is a very small isopod (average length 2.19mm) which is found in association with corals such as *Stylophora* and on coral rock.

Cirolana corrugis has a similar habitat on hard substrata in shallow or inter-tidal areas, frequently in association with the coral *Stylophora.*

Cirolana fishelsoni is about 3 to 4mm long. It lives in a wider range of habitats than the other two species of this genus listed above since it has been found in sandy and muddy areas and among *Halimeda* beds.

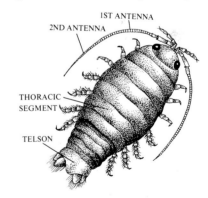

Figure 24: Generalized Isopod

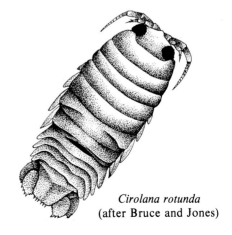

Cirolana rotunda
(after Bruce and Jones)

Idotea metallica

Lanocira latifrons (after Stebbing)

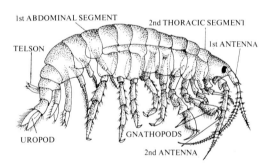

Figure 25: Generalized Amphipod
(after Barnes)

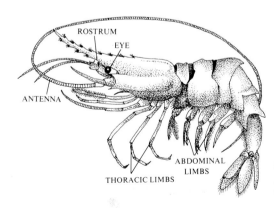

Figure 26: Generalized Decapod

Cirolana parva is present along the central Red Sea.

Cirolana theleceps is found in the Gulf of Suez and Aqaba and the Northern Red Sea where it occurs on coral rock.

Cirolana bovina occurs on rocky shores, around mid-tide level, where it lodges in crevices.

Excirolana orientalis is present on exposed beaches around mid-tidal level.

Tylos exiguus is a supra-littoral sandy beach species which has been found at Yenbo and along other beaches, usually coincident with the ghost crab *Ocypode saratan*.

Idotea metallica is a cosmopolitan isopod which occurs on sea-weeds.

Sphaeroma walkeri occurs in crevices on rocky shores around mid-tide level. It was originally described from South Africa where it is reported to be an estuarine species occurring in the shallow subtidal zone.

Lanocira latifrons was originally described from material collected in the Sudanese Red Sea. It is relatively common in crevices at mid-tidal level.

Exosphaeroma reticulatum occurs in the eulittoral zone on relatively exposed shore-lines.

Order: Amphipoda

Amphipods are readily distinguished by virtue of the fact that they are small, laterally compressed crustaceans which lack a carapace (figure 25). Abdominal appendages are modified for swimming or walking, and in those forms which live near the tide-line these appendages are particularly well developed and give rise to the name "sand-hoppers". In contrast to the copepods, there are relatively few planktonic amphipoda in the Red Sea.

Recorded species include: *Ampelisca sp.; Dexamanidae sp.; Prophliantinae sp.; Sphaerophthalmus sp.; Haustoriidae sp.; Urothoe sp.; Leucothoe sp.; Oedicerotidae; Synopia sp.; Rhabdosoma whitei; Oxycephalus clausi; O. erythraeus; Phronima sedentaria; P. atlantica* var. *solitaria; Platyscelus inermis.*

A common inter-tidal amphipod frequently found on sandy beaches together with ghost crabs *(Ocypode saratan)* and sand crabs *(Dotilla sulcata)* is *Talorchestia martensi.*

Superorder: Eucarida

Order: Decapoda

This important order of crustaceans includes most of the commercially important species such as shrimps, prawns, lobsters and crabs. The head and thorax are dorsally fused and covered by a carapace which extends down the sides of the body and encloses the gills. The first three thoracic appendages are modified as mouth parts while the remaining five pairs are leg-like and give rise to the name Decapoda ("ten-legged"). Decapod crustaceans are present in nearly all Red Sea habitats and they show a very wide range of adaptive radiation. Main characteristics are indicated in figure 26.

Suborder: Natantia: Shrimps and Prawns

The suborder Natantia is split into three infra-orders: *Penaeidea, Caridea* and *Stenopodidea.*

Infraorder: Penaeidea

Includes the families Penaeidae and Sergestidae. The former includes the commercially important shrimps: *Penaeus monodon; P. indicus; P. semisulcatus; P. latisulcatus; P. japonicus; Metapenaeus monoceros* and *M. stebbingi. Trachypenaeus curvirostris* is also present in small numbers in trawl catches. They are caught by cast-netting around mangrove areas and by shallow water trawling at night-time.

Infraorder: Caridea

Nearly all coral-reef associated shrimps belong to the Caridea. These are separated from Penaeid prawns by the fact that their third thoracic "legs" (pereiopods) are not clawed (chelate) in Caridea whereas they are in Penaeida.

FAMILY: PALAEMONIDAE

This is a large family of shrimps in which the second pair of chelipeds are usually longer than the first. Their rostrum is laterally compressed, serrated and usually quite long. While many are free-living, a considerable number of species live commensally with a wide variety of coral-reef organisms including corals, anemones, sponges, molluscs, echinoderms and even in association with fish. Their forms frequently raise the interest of SCUBA divers since they exhibit many interesting colour forms and behavioural characteristics. The family includes the familiar "common prawn" *Palaemon serratus* of European waters but of most interest to us are those coral-reef associated shrimps known as Pontoniinid shrimps.

There are at least 45 species of the subfamily Pontoniinae shrimps recorded from the Red Sea. They exhibit a wide range of forms, each adapted to a particular mode of life, and their morphological modifications parallel those occurring in other shrimp families (e.g. Alpheidae; Gnathophyllidae).

A list of recorded species is given in Table VII. Basic structure is typified by *Palaemonella elegans* (see figure 27).

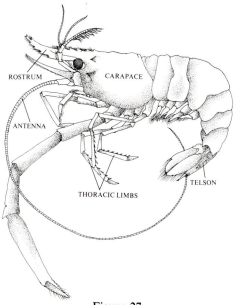

Figure 27:
Basic structure of Pontonine shrimps typified by *Palaemonella elegans*.

Table VII
SUB FAMILY PONTONIINAE **Shrimp Fauna of the Red Sea***

Anchistioides compressus	*Philarius gerlachei*
Anchistus custos, A. miersi	*P. imperialis*
Conchodytes biunguiculatus	*Pontoniopsis comanthi*
C. meleagrinae, C. tridacnae	*Periclimenes djiboutensis*
Coralliocaris graminea	*P. agag, P. brevicarpalis*
C. macrophthalma, C. nudirostris	*P. calmani, P. consobrinus*
C. superba, C. venusta	*P. diversipes, P. edwardsi*
Hamodactyloides incompletus	*P. elegans, P. ensifrons*
H. aquabai	*P. grandis, P. holthuisi*
Harpilopsis beaupresii	*P. imperator, P. kempi, P. longicarpus*
H. depressa	*P. longirostris, P. lutescens*
Jocaste lucinda	*P. ornatus, P. petitthouarsi*
Palaemonella rotumana	*P. soror, P. tenuipes, P. tenuis*
P. tenuipes	*Thaumastocaris streptopus*
Paratypon siebenrocki	

*After Bruce and Svoboda, 1983.

F. JACK JACKSON

Cleaner shrimp: *Periclimenes* sp. with Moray

Periclimenes soror

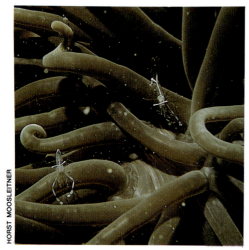

HORST MOOSLEITNER

Of these 75% are commensals of other marine organisms while only 16% are free living and one species *(Periclimenes tenuipes)* is a facultative commensal. Coelenterates are the most important hosts and 21 species (46%) are commensals of these. Fourteen are associated with scleractinian corals, five with anemones, two with soft corals and one with Hydroida. In addition to the Coelenterates, six species live with molluscs, three with echinoderms and one with tunicates.

Periclimenes species living with sea anemones in the Red Sea include: *P. tenuipes* (also found disassociated from anemones); *P. brevicarpalis; P. ornatus; P. longicarpus* and *P. holthuisi.*

Periclimenes kempi is found with the soft coral *Sarcophyton elegans.* Other species occur in association with black corals and sponges and the research of these is not yet complete. The reddish brown shrimp — *Periclimenes soror,* lives among the spines of the Crown of Thorns starfish — *Acanthaster planci.* It has a pale stripe along its body which mimics the form of a spine.

An interesting relationship is that between *Periclimenes imperator* and the nudibranch *Hexabranchus sanguineus.* The shrimp lives on the surface of the sea-slug, feeding on particles adhering to its mucus surface. The shrimp's red colour merges particularly well with that of the host and it only leaves the slug if it dies. It also occurs on sea cucumbers such as *Stichopus* and *Synapta* and on the sea-star *Gomophia egyptiaca.* Other associations are indicated in Table VIII.

The genus *Palaemonella* includes shrimps which show less specialisation with regard to hosts and many species are free-living active predators.

It is often overlooked that pontoniinid shrimps are capable of producing significant sounds and they even compete with the true snapping shrimps (Alpheidae) in this regard. *Coralliocaris graminea* is probably one of the best sound producers present and unlike alpheids they use both chelae and second walking legs to create sound.

Table VIII	
ASSOCIATIONS BETWEEN SOME COMMON RED SEA **Pontoniinae Shrimps and Various Hosts***	
Shrimp Species	**Associated Hosts**
1. *Periclimenes brevicarpalis*	*Cryptodendrum adhaesivum*
2. *P. soror*	*Acanthaster planci, Gomophia egyptiaca*
3. *P. imperator*	*Hexabranchus sanguineus, Gomophia egyptiaca, Stichopus & Synapta*
4. *P. holthuisi*	*Cassiopea andromeda*
5. *P. longicarpus*	*Gyrostoma helianthus, Megalactis hemprichii*
6. *P. ornatus*	*Gyrostoma quadricolor*
7. *P. tenuipes*	*Radianthus kosseirensis*
8. *Conchodytes meleagrinae*	*Pinctada margaritifera*
9. *C. biunguiculatus*	*Pinna sp.*
10. *Paratypon siebenrocki*	*Acropora sp.*
11. *Coralliocaris spp.*	Coral species
12. *Anchistus spp.*	Mollusca
13. *Pontoniopsis spp.*	Echinoderms

*After Bruce (1976).   *Periclimenes longicarpus*

HORST MOOSLEITNER

Periclimenes sp. on anemone

FAMILY: GNATHOPHYLLIDAE: Gnathophyllid Shrimps

Included in this family is the genus *Hymenocera* which has strongly developed mouth parts and whose members predate on starfish. *Hymenocera pictus* is one of the few predators on the Crown of Thorns starfish, *Acanthaster planci*.

FAMILY: ALPHEIDAE: Snapping Shrimps

Alpheid shrimps have attracted the attention of marine biologists and divers as a result of their habit of burrowing in the sea-bed and living in association with various species of gobies. The association between the two partners is a mutually beneficial one since the fish use the excavated burrow as a temporary refuge during daytime and as a permanent resting place at night. The shrimps have poor vision and they depend upon the goby for an early warning alarm system to protect against predators. There are at least 95 species of Alpheid shrimp reported from the Red Sea. The common names "snapping shrimps" or "pistol shrimps" result from their ability to make a cracking sound by manipulating their pincers. The sound is thought to frighten predators and it has been suggested that the rush of water which is created in the process of "snapping" may be used to stun their prey.

One common species living in shallow water burrows is *Alpheus djeddensis* which usually has one of several species of gobies associated with it (including *Lotilia graciliosa*, *Cryptocentrus caeruleopunctatus* and *C. sungami*). A more conspicuous species is *Alpheus lottini* (also recorded as "*A. sublucanus*") which is a commensal on *Stylophora pistillata* and *Pocillopora* corals. It usually has a bright orange-red ground colour with deeper red mottling and may have a darker red longitudinal stripe dorsally. *Aretopsis amabilis* (also known as *A. aegyptiaca*) associates with hermit crabs such as *Dardanus lagopodes*, sharing the same gastropod shell for shelter and probably feeding on faeces from the hermit crab.

Alpheus frontalis is found living in tubes attached to stones (sometimes corals such as *Stylophora pistillata*). It makes the tubes from the alga *Oscillaria*.

A list of Red Sea Alpheids is given in Table IX, and a generalized illustration is shown in figure 28.

Alpheus djeddensis with Goby

Table IX

RED SEA ALPHEIDAE*

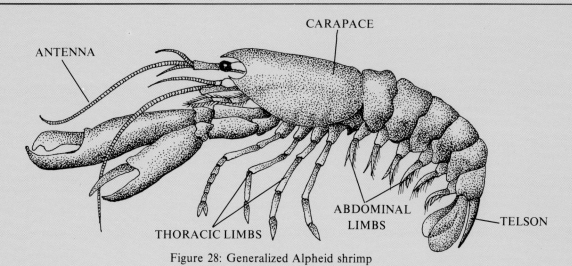

Figure 28: Generalized Alpheid shrimp

Alpheopsis equalis	*Alpheus macroskeles*	*Automate dolichognatha*
Alpheus alcyone	*A. maindroni*	*Betaeopsis indica*
A. alpheopsides	*A. malleodigitus*	*Racilius compressus*
A. barbatus	*A. microstylus*	*Salmoneus brevirostris*
A. bisincisus	*A. miersi*	*Salmoneus cristatus*
A. bucephaloides	*A. obesomanus*	*Salmoneus serratidigitus*
A. bucephalus	*A. pachychirus*	*Synalpheus ancistrorhynchus*
A. clypeatus	*A. pacificus*	*S. charon*
A. collumianus	*A. paracrinitus*	*S. coutierei*
A. crinitus	*A. pareuchirus*	*S. demani*
A. deuteropus	*A. parvirostris*	*S. fossor*
A. diadema	*A. rapacida*	*S. gracilirostris*
A. distinguendus	*A. rapax*	*S. hastilicrassus*
A. djeddensis	*A. serenei*	*S. heroni*
A. edamensis	*A. splendidus*	*S. mushaensis*
A. edwardsii	*A. spongiarum*	*S. neomeris*
A. ehlersii	*A. strenuus*	*S. nilandensis*
A. euchirus	*A. tenuicarpus*	*S. pachymeris*
A. euphrosyne euphrosyne	*Amphibetaeus jousseaumei*	*S. paradoxus*
A. fasciatus	*Aretopsis amabilis*	*S. paraneomeris*
A. frontalis	*Athanas areteformis*	*S. paulsoni*
A. gracilis	*Athanas crosslandi*	*S. paulsoni liminaris*
A. gracilipes	*Athanas dimorphus*	*S. physocheles*
A. hippothoe	*Athanas djiboutensis*	*S. quinquedens*
A. hululensis	*Athanas dorsalis*	*S. savigny, S. sladeni*
A. lanceloti	*Athanas ghardaqensis*	*S. spongicola*
A. lanceostylus	*Athanas indicus*	*S. streptodactylus*
A. leptochirus	*Athanas marshallensis*	*S. triacanthus*
A. leviusculus leviusculus	*Athanas minikoensis*	*S. tricuspidatus*
A. lobidens	*Athanas monoceros*	*S. trispinosus*
A. lottini	*Athanas sibogae*	*S. triunguiculatus*
A. macrodactylus	*Athanopsis brevirostris*	*S. tumidomanus*
	Athanopsis platyrhynchus	

*After Banner and Banner, 1981.

Hump-backed shrimp: *Saron marmoratus*

Cleaner shrimp: *Stenopus hispidus*

FAMILY: HIPPOLYTIDAE: Hump-backed Shrimps

These shrimps are characterized by a conspicuous hump on the abdomen. They are mostly free-living and are found under coral boulders or in other sheltered crevices, but some species live in association with anemones or sponges. Red Sea representatives include *Saron marmoratus,* which the author has found living on dead coral boulders forming the causeway to Suakin island, and is relatively common around shallow rocks in sheltered areas. In life the species has a colour pattern which makes it appear to be covered with anemones.
Thor amboinensis is a small hippolytid shrimp which is often found associated with large sea-anemones such as *Stoichactis gigas,* corals and hydroids. It has a dark olive brown body with numerous large pale opalescent patches. *Hippolysmata grabhami* is found among coral rubble in the fore-reef. It has been observed to clean fish.

Infraorder: Stenopodidea

This infraorder of the Natantia includes cleaner shrimps of the family *Stenopodidae,* which is represented in the Red Sea by the well-known and cosmopolitan banded coral shrimp *Stenopus hispidus.* This has its third pair of thoracic legs (chelipeds) much longer than the first two pairs. The shrimp rests in crevices or out in the open at its "cleaning station" and waves its long white antennae to attract fish or other crustaceans. When these approach, the shrimp first touches them with its antennae and then their "clients" become quite passive and allow the *Stenopus* to work over their body surfaces and to search their gills for parasites. The much photographed shrimp is unmistakable since it has a conspicuous red and white banding. They are usually found in pairs and these partnerships generally last for the entire life of the shrimps.

Suborder: Reptantia

Infraorder: Palinura: Spiny Lobsters

FAMILY: PALINURIDAE

There are three species of spiny lobsters belonging to the family Palinuridae known to occur in the Red Sea.

Panulirus pencillatus is the most common spiny lobster present in the region and is fished on exposed reefs close to deep water, such as occurs around some of the offshore islands and along sections of fringing reef. During summer months they move into relatively shallow water but in winter they probably migrate to deeper water. There is some variability in colour form, but all specimens have longitudinal pale bands on their legs. A measure of its importance and long history is provided by the fact that this species is illustrated on wall decorations of the ancient Egyptian temple at Deir-el-Bahari in a motif which depicts Queen Hatsheput's expeditions to Punt around 3,500 years ago!

Panulirus versicolor is also known as the "painted crayfish" since it is strongly patterned with dark and pale bands across the tail region. They are frequently observed by divers since, although during daytime they remain hidden in crevices among corals, their brilliantly white antennal flagella protrude and give the clue to their presence.

Panulirus ornatus is a relatively uncommon spiny lobster in the Red Sea and is mainly distributed in the Indo-Pacific where its populations may reach commercial significance. It is a species which seems to prefer relatively turbid water and it appears to be mainly confined to those areas of the Southern Red Sea where such conditions prevail, such as around Massawa and in areas of the Farasan bank.

FAMILY: SCYLLARIDAE: Slipper Lobsters

There are three genera of this family in the Red Sea.

Thenus orientalis: Unlike spiny lobsters, the slipper lobster *T. orientalis* lives on flat sandy or muddy areas and is often found close to the base of reef drop-offs. In parts of the Southern Red Sea, such as around Massawa, this species is fished by bottom trawling or by shallow water spear-fishing.

The genus *Scyllarus* is represented by: *Scyllarus ragosus; S. gibberosus; S. lewinsohni; S. pumilus.*

Scyllarides tridacnophaga is an unusual slipper lobster found in the Gulf of Aqaba (and probably also in the Red Sea proper), where it occurs in shallow water among coral boulders. As its name suggests, it feeds on large bivalves. The related species, *Scyllarides haanii,* is usually found close to the base of the main reef face. It may be distinguished from *S. tridacnophaga* since the carapace of *S. haanii* is widest in its posterior half and the cervical constriction is much more distinct than in the former species.

FAMILY: NEPHROPIDAE: Rock Lobsters

Enoplometopus occidentalis is a bright red colour with white spots on a hairy body. It occurs in deep water and is sometimes caught in deep trawls.

Infraorder: Astacura

These have a cylindrical carapace and the first three walking legs end in pincers, with the first pair of legs being much larger than the others. They possess a generally long narrow form with an elongate straight abdomen ending in a broad tail fan.

Represented species include: *Callianassa bouvieri* which is found burrowing in shallow sandy bays; *C. madagassa* and *C. novaebritanniae* which occur in sediment around mangroves; *Upogebia rhadames* which is often found in shallow water such as close to a pier in Suakin harbour.

Panulirus versicolor

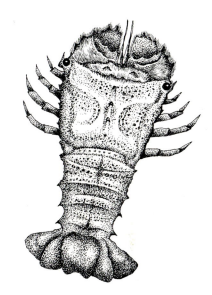

Thenus orientalis (after Ben-Tuvia)

Clam-eating lobster: *Scyllarides tridacnophaga*

Figure 29: *Dardanus tinctor* with *Calliactis* anemones on shell

Dardanus tinctor with *Calliactis* anemones on shell

Hermit crab: *Coenobita scaevola*

Infraorder: Anomura
Squat Lobsters; Hermit Crabs & Porcelain Crabs

These are walking decapods which possess pincers at the tips of the first pair of walking legs and have their fifth (and sometimes fourth) pair of legs reduced in size so that they are frequently obscured by the carapace. There are seven families which are discussed below.

FAMILY: DIOGENIDAE: Spotted Hermit Crabs

Perhaps the most widespread species is *Paguristes jousseaumei* which occurs on shallow reefs throughout the Red Sea. *Dardanus tinctor* is found among corals and has the habit of encouraging *Calliactis* anemones to settle on its gastropod shell (Figure 29). The related species *D. lagopodes* has a stranger association, sharing its gastropod shell — usually that of a stromb *(Strombus tricornis)* or a cone shell — with the alpheid shrimp *Aretopsis amabilis (= A. aegyptica)*. *Diogenes avarus* lives in sandy areas of the eulittoral. They bury into the sand as the tide ebbs and become active again when the tide floods back over the beach. Many of these crabs use the gastropod *Nassa arcularia* for their shelter. Several species, such as *Clibanarius longitarsus* are typically found among mangroves. A complete list of recorded species is given in Table X.

Table X	
RED SEA HERMIT CRABS	
DIOGENIDAE	**PAGURIDAE**
Paguristes jousseaumei	*Pylopaguropsis magnimanus*
P. perspicax, P. calvus	*Pagurus cavicarpus*
Paguristes sp.	*P. hirtimanus*
Clibanarius longitarsus	*P. boninensis*
C. striolatus	*P. cf. prideauxi*
C. signatus, C. carnifex	*Nematopagurus squamichelis*
C. virescens	*N. muricatus*
Dardanus tinctor	*N. diadema*
D. lagopodes	*Catopagurus ensifer*
D. woodmasoni	*Cestopagurus coutieri*
Diogenes avarus	*C. pectinipes*
D. costatus, D. gardineri	*C. helleri*
D. denticulatus	*Anapagurus bonnieri*
Calcinus latens, C. rosaceus	
Troglopagurus jousseaumei	**COENOBITIDAE**
Trizopagurus strigatus	*Coenobita scaevola*
T. shebae	

J. DAVID GEORGE

HORST MOOSLEITNER

FAMILY: PAGURIDAE: Hermit Crabs

These have a large, soft, spirally twisted abdomen which is protected by the shell of a dead gastropod which the crab carries around with it. One of the front claws is often larger than the other and is used to close off the entrance of the gastropod shell when the mollusc withdraws into it for protection. Hermit crabs are generally scavengers which are found in a wide range of habitats, especially among coral rubble on the reef flat. At least thirteen species are recorded from the Red Sea (Table X).

FAMILY: COENOBITIDAE: Land Hermit Crabs

These are similar in appearance to Pagurid hermit crabs but they have gill chambers which are able to utilise atmospheric oxygen and some species can remain out of water for long periods. Many species live in burrows or rest in shaded areas among coastal vegetation during daytime and then emerge at night to scavenge close to high water. The family is represented in the Red Sea by the ubiquitous species *Coenobita scaevola* which can be very abundant above sea level on beaches along mainland coasts or on offshore islands. They frequently occur together with ghost crabs and whole "armies" of them move along the shore at night when they seem to be guided towards dead fish or other food by their sense of smell. Their preferred "homes" seem to be shells of *Nerita*, *Cerithium* and small species of *Strombus*. At high tide (in those areas where a significant diurnal tide occurs) they dig into the sand just above high water mark. Their main breeding season is in spring but it can also occur in early summer or in autumn.

Coenobita scaevola showing tracks in sand

Hermit crab: *Dardanus lagopodes*

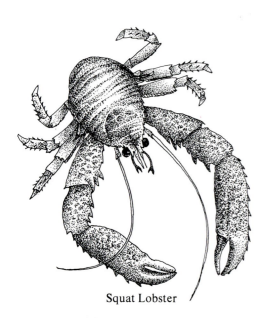

Squat Lobster

Above: Porcelain crab, *Petrolisthes rufescens;*
opposite, top: Box crab, *Calappa hepatica;*
bottom: Ghost crab: *Ocypode saratan.*

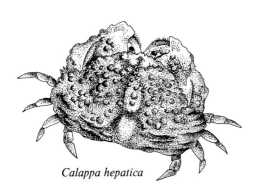

Calappa hepatica

FAMILY: GALATHEIDAE: Squat Lobsters

These have a more or less oval carapace with a sharp rostrum. They are small, lobster like crustaceans with long chelipeds normally extended in front of the animal. The abdomen is bent under the carapace so that their bodies appear to be very short. Several species are confined to deep water and some are commensals on different organisms, notably corals and crinoids.

Represented species are: *Galathea aegyptiaca; G. brevimana; G. longimana; G. orientalis; G. aculeata; G. gardineri; G. affinis; G. pusilla; G. humilis; G. genkai; G. elegans; Munida roshanei; M. japonica; M. gracilis; Bathymunida polae.*

Associations to watch for when diving are those of *Galathea genkai* with the crinoid *Heterometra savignii; G. elegans* also with this crinoid as well as *Lamprometra klunzingeri* and *Capillaster multiradius;* and that of *G. humilis* in *Seriatopora* and *Pocillopora* corals.

FAMILY: PORCELLANIDAE: Porcelain Crabs

Although these are superficially similar to crabs they are in fact more closely related to squat lobsters. They have long antennae and prominent chelipeds. Their tailed abdomen is normally tucked under the carapace. They are usually small "crabs" and while some are free-living and can be found among coral rubble, others live in association with a variety of other animals, notably anemones. They feed on suspended matter.

Represented species include: *Petrolisthes boscii; P. moluccensis; P. rufescens; P. leptocheles; P. tomentosus; P.. carinipes; P. pubescens; P. ornatus; P. virgatus; Pachycheles natalensis; Pachycheles sculptus; Pisidia inaequalis; P. serratifrons; Polyonyx biunguiculatus; Polyonx triunguiculatus; P. denticulatus; Polyonyx pygmaeus; P. pugilatus; P. pedalis; P. suluensis; Neopetrolisthes maculatus; Porcellana serratifrons; P. inaequalis.*

Of these, by far the most common species is *Petrolisthes leptocheles,* which occurs in a wide range of shallow water habitats and has been found living in association with the sponge *Spirastrella inconstans. Polyonyx pygmaeus* is a particularly small form (carapace length less than 5mm) which is found in association with the alcyonacean *Tubipora musica* and scleractinian corals of the genera *Galaxea, Acropora* and others.

FAMILY: HIPPIDAE: Mole Crabs

These have an almost cylindrical shape formed by a long carapace and an abdomen which is tucked under this. They are burrowing species which spend most of the time under sand or gravel in search of food. They can swim backwards and when disturbed they burrow rapidly into the sediment with legs which are specialised for digging.

Hippa picta and *H. celaeno* are both common on the lower shores and shallow sandy areas of the Red Sea. A third species, *Emerita emeritus* has also been recorded.

FAMILY: ALBUNEIDAE

Two species of this family have been recorded, i.e. *Albunea steinitzi* which lives on coral-sand in shallow lagoons, often associated with the burrowing hemichordate, *Ptychodera flava,* and *A. thurstoni* which is much less common.

Infraorder: Brachyura: True Crabs

The Brachyura includes all the true crabs and is split into around twenty-five families. Red Sea crabs are abundant in both numbers and in a variety of species. While some species such as ghost crabs make their presence known by their conspicuous burrow mounds, others are well camouflaged and remain hidden from all except the most determined biologists. Physical characteristics of true crabs include a dorso-ventrally flattened body and a carapace which is fused with a ventral plate. First legs have pincers which may be differentially developed. The reduced abdomen is tucked under the carapace.

Dromia dehaani

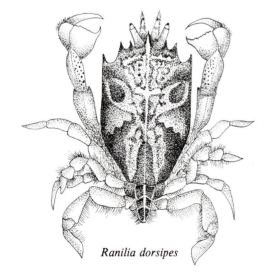

Ranilia dorsipes

FAMILY: DROMIIDAE: Sponge Crabs

These have domed furry carapaces which they usually cover with live sponges and ascidians. The crabs actually select pieces of sponge and then hold them in position with their rear thoracic legs.

Represented species are as follows: *Dromia dehaani; Dromidia unidentata; Cryptodromia hilgendorfi; C. bullifera; C. canaliculata; C. granulata; C. gilesii; C. globosa; Petalomera sp.; Pseudodromia caphyraeformis.* They are frequently found among coral rubble in shallow to moderate depths.

FAMILY: CALAPPIDAE: Box Crabs

Box crabs can fold their legs into the body in such a way as to create the neat form of a box. Their right cheliped carries a toothed claw used to open gastropod shells upon which they prey. In addition to attacking the live gastropods they may also eat hermit crabs occupying dead gastropod shells.

A common species in shallow water is *Calappa hepatica.* Others include *C. philargius; C. gallus; C. dumortieri; Matuta lunaris* and *M. banksi.*

FAMILY: RANINIDAE: Spanner Crabs

These have an elongated carapace but it does not cover the abdomen. They are named "spanner crabs" because the tips of the first thoracic legs have spanner-shaped pincers in which the movable claw is at right angles to the flattened end of the movable joint. They live in coral sand, and usually lie with only the front of their bodies protruding from the sand.

Represented species include: *Ranilia dorsipes* (also referred to as *Notopus dorsipes*) and *Cosmonotus grayi.*

FAMILY: LEUCOSIIDAE: Pebble Crabs
These are distinguished by a globular carapace which is pointed anteriorly. They have well-developed chelipeds and are found among coralline rubble in shallow water. Represented species are listed in Table XI.

FAMILY: DORIPPIDAE
Represented in the Red Sea by *Dorippe frascone*.

FAMILY: MAJIDAE: Spider Crabs
These have a more or less triangular-shaped carapace and long, slender walking legs which make them resemble a spider. They are frequently well-camouflaged by attached organisms such as algae or hydroids and their legs and carapace have hair-like projections which assist in the attachment of these sedentary creatures. Red Sea spider crabs are listed in Table XII.

Nut crab: *Leucosia signata*

Table XI
RED SEA PEBBLE CRABS: LEUCOSIIDAE

SUBFAMILY

EBALIINAE	*PHILYRINAE*	*LEUCOSIINAE*
Nucia speciosa	*Arcania septemspinosa*	*Cryptocnemus tuberosus*
N. pulchella	*A. quinquespinosa*	*Leucosia signata*
Ebalia granulata	*Iphiculus spongiosus*	*L. anatum*
E. orientalis	*Ixa cylindricus*	*L. elata*
E. abdominalis	*I. inermis*	*L. corallicola*
E. lacertosa	*Myra fugax*	*L. hilaris*
E. nobilii	*M. affinis*	
Oreophorus horridus	*M. kessleri*	
	M. brevimana	
	M. pentacantha	
	Nursia rubifera	
	N. jousseaumei	
	N. dimorpha	
	Nursilia dentata	
	Pariphiculus coronatus	
	Philyra scabriuscula	
	P. variegata	
	P. platychira	
	P. granigera	
	Pseudophilyra tridentata	

Philyra platychira

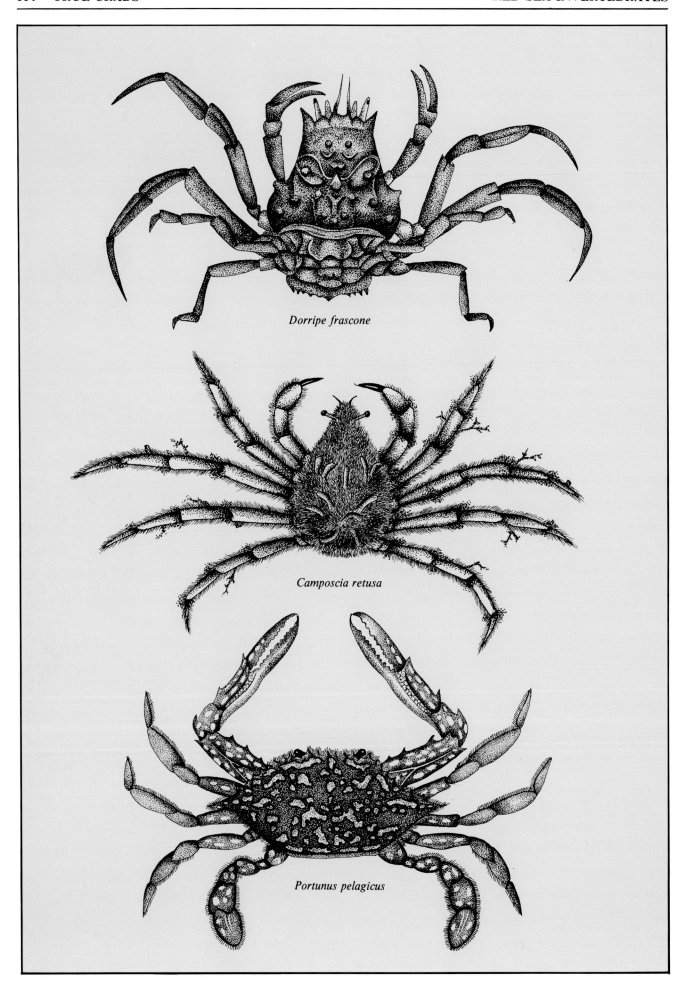

Dorripe frascone

Camposcia retusa

Portunus pelagicus

Table XII

RED SEA SPIDER CRABS (MAJIDAE)*

SPECIES	GEOGRAPHIC RANGE			RED SEA HABITAT	
	Indo-West Pacific	Indian Ocean	Red Sea	Recorded Depths 10m or less	More than 10m
Acanthonyx elongatus		x		x	
A. dentatus		x		x	
Achaeus brevifalcatus					
A. brevirostris					
A. erythraeus					
A. lorina					
Aepinus indicus	x				x
Camposcia retusa	x			x	
Cyclax spinicinctus	x			x	
C. suborbicularis	x				
Cyphocarcinus minutus		x		x	
Entomonyx spinosus	x				x
Eurynome stimpsoni		x		x	
E. verhoeffi			x		x
Hoplophrys oatesii	x				x
Huenia proteus	x				
Hyastenus convexus	x				x
H. diacanthus	x				x
H. elongatus	x				x
H. hilgendorfi	x				x
H. inermis		x			x
H. spinosus	x				x
Inachoides dolichorhynchus		x			x
Lambrachaeus ramifer	x				x
Maxioides spinigera (var. inermis)		x			x
Menaethiops contiguicornis			x		
M. dubia			x	x	
M. ninni			x	x	
M. nodulosa		x		x	
M. monoceros	x			x	
Micippa platipes	x			x	x
M. thalia	x				
Oncinopus neptunus	x				x
Ophthalmus cervicornis	x				
O. curvirostris		x			x
O. longispinus			x		x
Perinia tumida	x			x	
Phalangipus hystrix	x				x
Pseudomicippe nodosa		x			
Schizophrys aspera	x			x	
Simocarcinus pyramidatus		x		x	
S. simplex	x			x	
Stilbognathus erythraeus		x		x	
S. soikai			x		
Tylocarcinus styx	x			x	
Xenocarcinus tuberculatus	x				x

* (after Griffin & Tranter, revised, 1974)

HORST MOOSLEITNER

Portunid crab: *Lissocarcinus orbicularis:* symbiotic on surface of sea anemones.

Lissocarcinus polybioides

FAMILY: PORTUNIDAE: Swimming Crabs

The Portunid crabs number among their members the common European shore-crab *Carcinus maenas* which is familiar to most people who have turned rocks or snorkelled along shore-lines of the Mediterranean or North Atlantic. It occurs in and around the Suez Canal and in some intertidal regions of the North Red Sea. The true swimming crabs belong to the sub-family Portuninae and they tend to have broad carapaces which may be armed with sharp elongate laterally protruding spines. Their fifth pair of legs are flattened to form paddles which they use for swimming and burying themselves in the sand. Some members of the family are large enough to provide a useful food source but crabs are not a favoured food in most countries bordering the Red Sea.

Portunus pelagicus is common in shallow harbours and bays along both coastlines.

Scylla serrata is readily identified by the regular serrations around its carapace. It lives in muddy areas, often in association with mangroves. A complete list of Red Sea Portunids is given in Table XIII.

Table XIII
PORTUNIDAE FROM RED SEA

Subfamily: **CARCININAE**	**Subfamily:** **PORTUNINAE**
Carcinus maenas	*Lupocyclus philippensis*
	Charybdis variegata, C. natator
	C. lucifera, C. erythrodactyla
Subfamily: **CATOPTRINAE**	*C. anisodon, C. orientalis*
	C. paucidentata, C. japonica
Carupa tenuipes	*C. helleri, C. hoplites*
	C. vadorum, C. heterodon
	C. obtusifrons, C. amboinensis
	C. hoplites var. longicollis
Subfamily: **CAPHYRINAE**	*Thalamita prymna, T. chaptali*
	T. poissoni, T. crenata, T. sima
Caphyra polita	*T. integra, T. picta, T. stimpsoni*
Lissocarcinus polybioides	*T. quadrilobata, T. granosimana*
	T. bandusia, T. demani
	Thalamitoides quadridens
Subfamily: **PODOPHTHALMINAE**	*Thalamitoides tridens spinigera*
	Portunus pelagicus
Podophthalmus vigil	*P. sanguinolentus, P. convexus*
	P. tenuipes, P. granulatus
	P. argentatus, P. longispinosus
	P. alcocki, P. arabicus
	P. orbitosinus
	Scylla serrata

Parthenope diacanthus

FAMILY: PARTHENOPIDAE: Parthenopids

These also have a triangular carapace but it is covered with protruberances. They have slender antennae and eyes which can be retracted into their orbits.

Represented species include: *Lambrus pelagicus; L. carinatus; L.
lamelliger; L. calappoides; L. hoplonotus; L. beaumonti; Thyrolambrus
leprosus; Heterocrypta investigatoris; H. petrosa; Daldorfia horrida; D.
acuta; Eumedonus pentagonus; E. zebra; E. granulosus; Ceratocarcinus
spinosus; Parthenope (Autacolambrus) diacanthus.*

FAMILY: XANTHIDAE: Dark Fingered Coral Crabs

The family Xanthidae is one of the best represented families of crabs
associated with Red Sea reef corals. There is a wide range of forms and
adaptations within the family but nearly all members have black tipped
thoracic legs. They have broad, fan-shaped carapaces with well-separated eyes. Colour patterns vary from brightly coloured to drab. They
may be highly poisonous to eat. Some species live as commensals of
corals and may show a preference for particular coral genera. For
example, *Trapezia spp.* occur in *Pocillopora* and *Stylophora* corals;
Tetralia spp. live with *Acropora* and *Stylophora* corals and *Quadrella* live
with soft corals and gorgonians. The boxer crab *Lybia leptocheilus* has
the remarkable habit of carrying small specimens of the anemone *Triactis producta* on its front claws which it waves in defence against any
would-be predators. *Pilumnus incanus* is locally common in shallow
rocky areas and is characterized by a hairy carapace somewhat reminiscent of a wooly coat.

Trapezia cymodoce

Actumnus hirsutissima can be very common among stones or jetty
pilings close to the fringing reef edge. Specimens are frequently infected
with the parasitic barnacle *Sacculina*.

A list of Red Sea Xanthidae (123) species is given in Table XIV.

Table XIV

XANTHIDAE FROM RED SEA

Actaea hirsutissima
A. rufopunctata rufopunctata
A. savignyi, A. tomentosa
A. calculosa, A. vermiculata
A. nodulosa, A. cavipes
A. speciosa, A. polyacantha
A. helleri, A. sabaea
A. retusa, A. remota

Banareia parvula
B. kraussi, B. armata
B. banareias, B. nobilii

Atergatis floridus
A. roseus, A. granulatus

Atergatopsis signatus
A. granulatus

Carpilius convexus

Chlorodiella nigra
C. cytherea
C. laevissima, C. bidentata

Cycloxanthops lineatus

Cymo andreossyi
C. melanodactylus
C. quadrilobatus

Etisus electra, E. anaglyptus
E. laevimanus, E. splendidus
E. paulsoni, E. demani

Euxanthus sculptilis

Halimede ochtodes

Hypocolpus diverticulatus

Lachnopodus subacutus

Leptodius exaratus
L. sanguineus, L. gracilis
L. crassimanus, L. voeltzkowi

Liocarpilodes integerrimus
L. armiger

Liomera rugata
L. bella, L. rugipes, L. rubra
L. edwardsi, L. lophopa

Lophozozymus pulchellus

Medaeus simplex
M. granulosus, M. noelensis

Neoliomera themisto
N. richtersi, N. nobilii

Phymodius monticulosus
P. nitidus, P. granulatus, P. drachi

Pilodius areolatus
P. pugil, P. spinipes

Platypodia granulosa
P. anaglypta, P. semigranosa
P. cristata, P. helleri

Pseudoactumnus pestae

Pseudoliomera granosimana
natalensis

Xanthias punctatus
X. latifrons, X. cumatodes

Zozymodes xanthoides, Z. cavipes

Zosimus aeneus

Subfamily: MENIPPINAE

Domecia hispida

Epixanthus frontalis, E. corrosus

Eriphia sebana smithi
E. scabricula

Globopilumnus calmani

Menippe rumphii

Pseudozius caystrus

Lydia tenax

Schaerozius nitidus

Subfamily: PILUMNINAE

Actumnus asper, A. setifer setifer,
A. miliaris

Eurycarcinus natalensis

Glabropilumnus laevimanus

Heteropilumnus fibriatus,
H. setosus, H. quadrispinosus,
H. trichophoroides H. lanuginosus

Heteropanope convexa

Lybia caestifera

Pilumnus vespertilio, P. forskali,
P. hirsutus, P. savignyi,
P. longicornis, P. propinquus

Pilumnopeus vauquelini, P. laevis,
P. pharaonicus

Parapilumnus quadridentatus

Planopilumnus spongiosus

Subfamily: TRAPEZIINAE

Tetralia glaberrima

Trapezia cymodoce cymodoce,
T. rufopunctata
T. ferruginea, T. digitalis,
T. guttata T. cymodoce intermedia

Carpilius convexus

Actaea rufopunctata

Pilumnus sp.

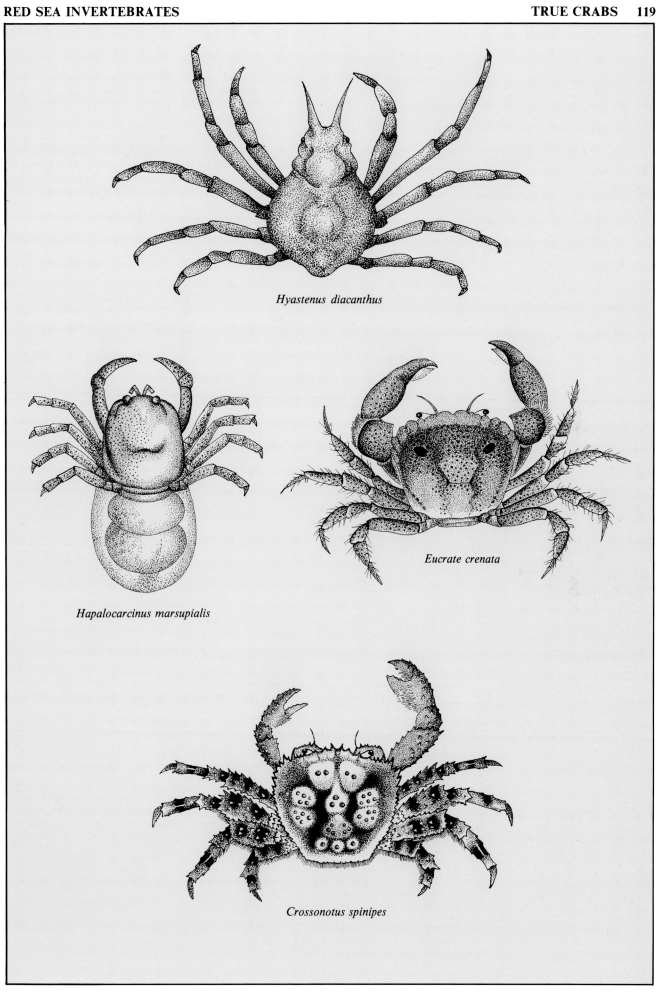

Hyastenus diacanthus

Hapalocarcinus marsupialis

Eucrate crenata

Crossonotus spinipes

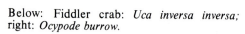

Pea crab

FAMILY: GONEPLACIDAE

Representatives include: *Carcinoplax longimanus; C. purpurea; Eucrate crenata; Litocheira integra; Notonyx vitreus; Paranotonyx curtipes; Typhlocarcinus rubidus; T. villosus; Xenophthalmodes moebii.*

FAMILY: PINNOTERIDAE: Pea Crabs

These usually live in close association with other organisms such as clam shells and their carapace is soft and membranous. Representatives include: *Duerckheimia carinipes; Ostracotheres tridacnae; O. affinis; O. cynthiae; Pinnoteres pernicolus; P. pectinicolus; P. purporeus; P. pilumnoides; P. maindroni; P. borradailei; P. coutieri; P. lutescens.*

FAMILY: HAPALOCARCINIDAE: Gall Crabs

These earn their common name from the protuberant bulges which their presence on live corals induces in the growth of the corals. Young crabs settle on corals such as *Stylophora, Pocillopora* or *Seriatopora* and the coral secretes a gall sac which encloses the crab, except for a very small opening kept clear by the crab. Thus entrapped within the coral, gall crabs enjoy protection from predators (except those which crunch up the coral) and they are able to feed on plankton which is collected on their hairy mouthparts. Red Sea species are *Hapalocarcinus marsupialis* and *Cryptochirus coralliodytes.*

FAMILY: CYMOPOLIIDAE

Represented in the Red Sea by *Manella spinipes.*

Below: Fiddler crab: *Uca inversa inversa;* right: *Ocypode burrow.*

HORST MOOSLEITNER

J. DAVID GEORGE

Below and opposite: Ghost crab: *Ocypode saratan*

JIM DORAN

FAMILY: OCYPODIDAE
Stalk-eyed Crabs; Ghost Crabs and Fiddler Crabs

Sandy shores along Red Sea coastlines are often marked by pyramid shaped mounds formed by burrowing ghost crabs. Males construct 15cms high sand mounds about half a metre from their burrows in order to attract females into the proximity of their burrows whereupon they entice them to enter by gesticulating with their large orange-yellow claws. During daytime the crabs tend to be hidden in their burrows but at night-time they emerge to scavenge for food along the tide-line. If one walks these beaches at night, it is possible to approach them quite closely. They have a somewhat ghostly appearance with eyes carried on long stalks.

By far the commonest ghost crab on sandy shores of the Red Sea is *Ocypode saratan. O. cordimana* is also present in the Gulf of Aqaba and along the western coast of the Red Sea. Other genera in the family include: *Uca, Dotilla, Macrophthalmus* and *Paracleistostoma.*

P. leachii is a Red Sea endemic crab which is found among the mangrove *Avicennia marina* in Eritrea and elsewhere in the Red Sea. The *Macrophthalmus* and Fiddler crabs *(Uca spp.)* occur in shallow muddy areas, often associated with mangroves. The sand crab, *Dotilla sulcata* is also common in shallow sandy bays.

A list of recorded species is given in Table XV.

Table XV
OCYPODIDAE FROM THE RED SEA

Ocypode saratan

O. cordimana

Uca lactea albimana

U. inversa inversa

Dotilla sulcata

Macrophthalmus telescopicus

M. grandidieri

M. graeffei

M. depressus

M. boscii

Paracleistosoma leachii

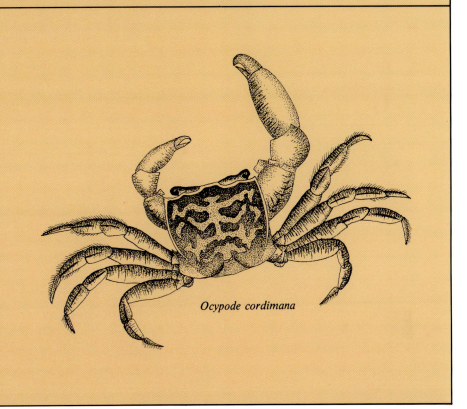

Ocypode cordimana

JIM DORAN

J. DAVID GEORGE

Swift-footed crab: *Grapsus albolineatus*

FAMILY: PALICIDAE

There are two genera and four species recorded from the Red Sea. They are as follows: *Palicus carinipes* (caught in trawls 25 to 85m); *P. elaniticus* (60-80m in Gulf of Aqaba); *P. whitei; Crossonotus spinipes.*

FAMILY: GRAPSIDAE: Shore Crabs

These possess somewhat angular carapaces and have powerful legs which have given rise in at least one instance to the common name "swift-footed" crabs. They scurry along shore-line rocks or among mangroves and can also occur in shallow beach rock crevices inside fringing reefs. They are generally difficult to catch since they escape into the deep crevices or holes when chased. Represented species are listed in Table XVI.

Of these, *Grapsus albolineatus* is one of the commonest species. It has a widespread distribution, occurring among rocks at or above the water-line and spends much of its time out of water. The smaller *Grapsus granulosus* appears to be endemic to the Red Sea and most recordings of it have been from northern regions. *Metopograpsus messor* has a wide distribution and is typically found in the littoral zone underneath stones in muddy areas. It is also present among mangroves.

Other species also found among the aerial roots of the mangrove *Avicennia* are *Metopograpsus thukuhar, Sesarma guttatum* and *Sarmatium crassum*. Indeed, these three species appear to be true mangrove crabs, not occurring away from mangroves. *Pachygrapsus minutus* is common in the Northern Red Sea and Gulf of Aqaba where it occurs on the reef-flat, often in quite large aggregations. It has been reported to be quite poisonous to eat. *Grapsus tenuicrustatus* like most grapsids, hides in holes or crevices during day time but emerges at night to forage along the shore, especially on low fossil cliffs which occur along both coastlines of the central Red Sea. *Geograpsus crinipes* has a similar habitat.

Table XVI

SHORE-CRABS: GRAPSIDAE FROM THE RED SEA

Grapsus tenuicrustatus	*Sarmatium crassum*
G. albolineatus	*Helice leachii*
G. granulosus	*Plagusia tuberculata*
Geograpsus crinipes	*Percnon planissimum*
Metopograpsus messor	
M. thukuhar	
Pachygrapsus minutus	
Iliograpsus paludicola	
Pseudograpsus elongatus	
Utica barbimana	
Sesarma guttatum	
Nanosesarma jousseaumei	

Grapsus albolineatus

FAMILY: GECARCINIDAE: Land Crabs

Whether one should include this family in an account of Red Sea invertebrates is perhaps debatable, for they are found on land which may be seasonally flooded with fresh water. They are able to use fresh water or sea water to keep their gills moist. They burrow into embankments and may occur in areas of fresh or brackish wells and salt marshes. *Cardisoma carnifex* is the main species of this family living on Red Sea coastal regions. A few specimens of *Gecarcoidea lalandii* have also been reported.

PHYLUM: CHELICERATA

Class: Pycnogonida: Sea-spiders

Spindle-legged, minute, spider-like creatures often found in association with encrusting fauna are likely to belong to the same phylum as land-spiders, i.e. Chelicerata, but to a different class, Pycnogonida. They are usually only a few millimetres in length.

FAMILY: NYMPHONIDAE

Nymphon maculatum is about 6mm long (including the proboscis). The head is half the total length of the body and the neck is elongate and slender. There is a short, stout proboscis. As is normal in this group, the male carries the eggs (see figure 30). This species was described from material collected from a navigational buoy floating in Port Sudan harbour and such well-encrusted habitats are typical for the species.

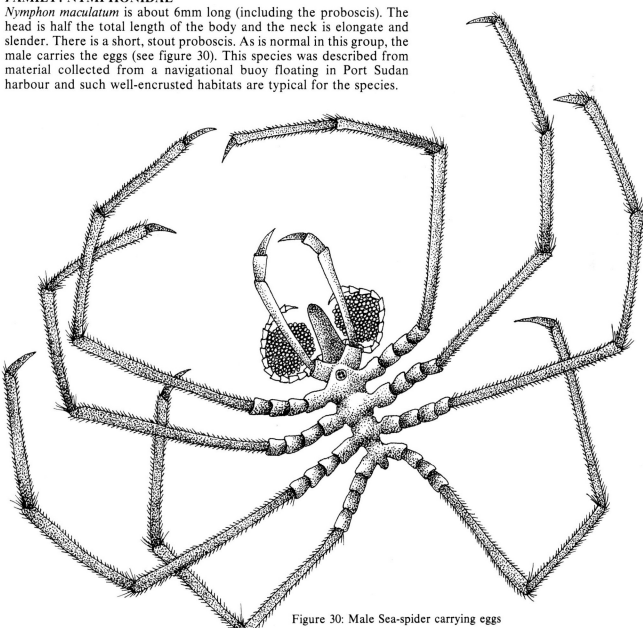

Figure 30: Male Sea-spider carrying eggs

J. DAVID GEORGE

7. MOLLUSCS

PHYLUM: MOLLUSCA

Class: Polyplacophora: Chitons

Chitons are found on dead coral rocks mainly in the intertidal and shallow reef zones. They have flattened bodies with eight overlapping transverse plates which are partially embedded in, and surrounded by, a relatively thick and fleshy girdle which may have spines or scales. The underside is dominated by a large muscular foot. The head lacks eyes or tentacles. Most species become active at night when they use a well-developed radula to scrape their algal food from rock surfaces.*

FAMILY: ISCHNOCHITONIDAE
Ischnochiton yerburi is commonly found attached to shaded surfaces of coral boulders in shallow water, especially between the shore and the reef platform. The girdle lacks spines or scales and does not extend over the plates.

FAMILY: CHITONIDAE
Members of this family possess a girdle which is usually heavily armed with spines or scales but which does not extend over the plates.

Tonicia suezensis is not confined to the Northern Red Sea although it is a common species there. It occurs under micro-atolls in the shallows and attached to loose rock.

Chiton olivaceus affinis is present throughout the Red Sea coastal waters where it is found attached to dead coral on the reef platform, and also lodged in crevices and on walls of caverns in the shallow fore-reef zone, just below the reef crest.

Acanthopleura haddoni

Below left: *Chiton olivaceus affinis;* below, *Acanthopleura haddoni.*

*Note: The author acknowledges assistance received in the preparation of this chapter by Dr. Horst Moosleitner.

J. DAVID GEORGE

Acanthopleura haddoni

Acanthopleura haddoni is common in very shallow water, for example, submerged beach-rock or on harbour walls, as well as dead coral boulders in the littoral zone. In the central Red Sea, where a true littoral zone is virtually non-existent (due to the very small tidal amplitude), an equivalent zone exists on exposed rocky surfaces such as jetties or reef markers where breakers continually splash the rock for a few centimeters above mean sea-level. This chiton is an active algal grazer which feeds upon the dense lawns of filamentous algae generally coating rock surfaces in the wave splash zone.

FAMILY: ACANTHOCHITONIDAE
Includes several large species. Shell plates have well developed teeth at their edges and they are partially covered by the fleshy girdle.

Acanthochiton penicillatus is mainly recorded from the Northern Red Sea and its twin gulfs, where it is relatively common on the undersides of coral boulders or micro-atolls in shallow water.

Class: Gastropoda: Gastropod Molluscs

Order: Archaeogastropoda
Primitive gastropods which include limpets, abalones and top-shells, together with some less well-known forms. The shell varies from spirally coiled to cone-shaped forms and is usually lined with mother-of-pearl.

FAMILY: SCISSURELLIDAE
These have a tiny, spirally coiled shell (-3mm) with an open slit in the outer lip of younger specimens, e.g. *Scissurella reticulata*.

FAMILY: HALIOTIDAE: Abalones and Ormers
Ear-shaped shells with a compact spiral at the apex and a series of perforations around the main whorl.

Sanhaliotis pustulata has an uneven outer surface and four raised holes around the periphery. It is a relatively common species occurring under boulders and on shaded surfaces of micro-atolls in shallow lagoons or in the littoral zone.

Sanhaliotis varia has five holes and cording. It may be common under coral heads and in crevices in shallow water.

Sanhaliotis pustulata

FAMILY: PATELLIDAE: Limpets

Cone-shaped shells which cling tightly to rocks in shallow water or in the littoral (or exposed splash) zone. There is no operculum and limpets escape predators by clamping themselves to the substrate with the aid of a large muscular foot which they also use for crawling across their substrate. They use a well developed radula for grazing on algae.

Cellana eucosmia is a relatively large endemic species, approximately 4.5cms in diameter, with a radial patterned shell (usually with nine main sectors). It occurs intertidally and in the central Red Sea in the exposed splash zone, on structures such as reef-markers, light-houses jetties or harbour walls.

Cellana eucosmia

FAMILY: FISSURELLIDAE: Key Hole Limpets

These generally have a hole at the apex of their cone-shaped shells but in members of the genus *Emarginula* have a slit on the front margin of the shell replacing the hole.

Hemitoma tricarinata is similar to members of the *Patellidae* but with a hole at the apex. It is found on fossil beach-rock and under boulders.

Diodora rueppelli is a relatively small shell (approx. 2.5cms) with a radiating pattern of dark and lighter segments. It has a hole at the apex.

Emarginula rugosa has a slit in front of its prominently ribbed shell which lacks a hole at the apex.

Scutus unguis has a flatter shell with concentric growth lines and a slight (easily overlooked) indentation at its front margin. It is found in sea-grass beds and grows up to about 5cms in length.

Diodora rueppelli

Emarginula rugosa

Scutus unguis

FAMILY: TROCHIDAE: Topshells

These are cone-shaped, spirally coiled shells which always possess a horny operculum, which is used to close the aperture when they are disturbed. Most have a nacreous layer on the inside of their shell and the larger species have been commercially collected for many years as a source of mother-of-pearl. They graze on algae growing on dead coral.

Trochus erythreus is a relatively small species which may be fairly common intertidally on coral rock, weeds and mangroves or similar habitats in shallow water to around 3m.

Tectus dentatus (previously regarded as a member of the *Trochus* genus) is a large trochus shell which occurs on the shallow reef-flat where its algal food is in good supply. It may also be found near the reef crest or on the slope.

Clanculus pharaonis is aptly named "Strawberry Topshell" since it has a red ground colour with a beaded black and white pattern. It is common on the underside of littoral or shallow rocks and among coral rubble. It has a relatively small shell (approximately 2cms).

Clanculus pharaonis

S. HALL

Trochus lugger: *Tectus dentatus*

JIM DORAN

Opposite: *Trochus (= Tectus) dentatus*
Left: *Turbo petholatus.*

FAMILY: TURBINIDAE: Turbans

Turbans resemble *Trochidae* but they have a calcareous operculum instead of a horny one. They are also shallow water herbivorous species.

Turbo petholatus: the Tapestry Turban.
It has a shiny shell and a "cat's eye" calcareous operculum which will be familiar to many readers. It lives on shallow reef stones.

FAMILY: STOMATELLIDAE: False Ear Shells

These resemble Haliotidae (Abalones) but there are no holes in their shell. They are also herbivores and are generally found in relatively shallow water.

Stomatia phymotis is a very fragile irregular shell which is found attached to littoral rocks and boulders.

FAMILY: PHASIANELLIDAE: Pheasant Shells

These are spirally coiled colourful shells.

Phasianella variegata is a common species on sea-grass beds.

FAMILY: NERITIDAE: Slipper Winkles

Members of this family are among the most tolerant gastropods to the extreme conditions of heat and dessication which exist in the Red Sea intertidal and supralittoral zones. They are capable of storing water to enable them to survive long periods of emersion.

Nerita undata quadricolor has a variable colour and a ridged surface. It is found on boulders or rocks in very shallow or littoral zones, especially on concrete ramparts of reef-markers and light-houses where it may be found in the splash zone a few centimetres above high water level.

Nerita albicilla also has a variably patterned, but more brightly coloured shell and a generally similar habitat to *Nerita undata quadricolor,* but tends to be hidden under stones during the day and emerges to graze upon the same stones at night.

Ecological studies on *Nerita* species in the Indian Ocean throw some interesting light on the behaviour of the same species which are found in the Red Sea. All species confined their foraging activities to the hours of darkness, generally at low-tide. *Nerita undata* is cryptically coloured and often visible during daytime, whereas *Nerita albicilla* is often brightly and variably coloured but almost always concealed in the day.

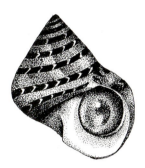

Turbo petholatus

Phasianella variegata

Nerita undata

J. DAVID GEORGE

Slipper winkles: *Nerita albicilla*

Littorina scabra

Order: Mesogastropoda

A large order of gastropods which exhibit a more advanced arrangement of the mantle cavity, i.e. a single gill with leaflets on only one side of the central shaft. The shell is usually spirally coiled and the aperture is closed by a horny operculum. The radula characteristically bears seven teeth in each transverse row. The order includes many familiar molluscs such as periwinkles, conchs and cowries. Most are bottom dwellers although a few free-swimming forms exist. Herbivores, detrital feeders and predatory species are represented.

FAMILY: LITTORINIDAE: Periwinkles

Periwinkles are small, spirally coiled gastropods which are particularly common in the intertidal zone. They have a horny operculum and the aperture of their shell is porcellaneous and smooth, with no siphonal canal. In the central Red Sea, where diurnal tides are greatly restricted, these forms may be found on rocky surfaces where exposure to waves creates a fairly consistent splash-zone a few centimetres above high tide level. Others live in association with mangroves.

Littorina scabra has a thin shell (approx. 4cms) of somewhat variable shape and colour but usually with brown axial (sometimes zig-zag) lines over a yellow ground colour. In the central Red Sea it is common on concrete ramparts of harbour and navigational structures and on fossil beachrock. It is also reported to favour attachment to mangrove trunks, particularly in the Southern Red Sea.

Nodilittorina millegrana also has a variable shell which may have a somewhat granular surface. It is quite common on concrete navigational structures and on fossil beachrock in very shallow or intertidal zones.

FAMILY: RISSOIDAE: Rissoids

These are tiny shells which may occur in very large numbers on certain algae and on sea-grasses. Examples include *Rissoina bouveri*. Twenty-two different species were recorded from reefs near Aqaba.

FAMILY: ARCHITECTONICIDAE: Sundials

These are flat to discoid shells in which the whorls are distinct and the umbilicus is so wide that the female keeps her egg capsules in it. They occur on seaweed or on sand and often under coral heads or in crevices in shallow water.

Heliacus variegatus is one of the most frequently found members of the family.

Heliacus variegatus

FAMILY: TURRITELLIDAE: Turret or Screw Shells

These have long slender pointed shells in which the whorls are rounded and have a relatively dull surface. The aperture is simple and lacks any special protuberances such as a siphonal canal. The operculum is flexible and horny.

Turritella maculata has a thickened shell (approx. 10cms) with a deep suture and prominent spiral ribs on each whorl; the colour is variable but generally creamy white with brown or purple patches or stripes or fine brown spiral lines. It may sometimes be plain white without any other markings. It is found in sand in the fore-reef zone, in between coral outcrops.

Turritella terebra is another relatively common species.

FAMILY: VERMETIDAE: Worm Shells

Worm-molluscs live embedded in the substrate which is often a living coral or the hydro-coral *Millepora*.

Dendropoma maximum is a dominant species of outer reef-flats in the Red Sea where it may reach densities of more than 22 specimens per m². It seems to prefer strong water movement and feeds by a mucus net which is spread by wave action over the reef surface. About once every quarter of an hour, the vermetid hauls in its mucus net together with any trapped food particles. Females brood egg capsules which are suspended by stalks from the roof of the shell. Each capsule contains up to 500 embryos and this brooding behaviour avoids any planktonic phase, thus enabling the young vermetids to maintain their position in their favoured zone of high wave energy.

Serpulorbis inopertus occurs on littoral rocks and boulders, on cemented reef platform and on similar hard surfaces in shallow water.

Turritella terebra

HORST MOOSLEITNER

Worm mollusc: Vermetid: *Dendropoma maximum* with mucus feeding net visible.

Planaxis sulcatus

Terebralia palustris

Rhinoclavis fasciatus

Cerithium erythraeonense

Epitonium aculeatum

Opposite: Panther cowrie: *Cypraea pantherina*

FAMILY: PLANAXIDAE: Cluster Winks

These earn their common name from their habit of clustering together in large numbers on rocky shores. They can be separated from Littorinids due to the presence of a distinct notch at the base of the columella.

Planaxis sulcatus occurs on stones near the water's edge or around intertidal pools. In the mid-Red Sea it is also a characteristic member of the "splash zone" community which is present on exposed concrete navigational structures such as jetties, reef markers and light-house foundations.

FAMILY: POTAMIDIDAE: Telescope Shells

Long, slender, spirally coiled, frequently drab shells which are often associated with rather brackish water. They superficially resemble Cerithids but the siphonal canal is *not* twisted to the rear or upturned. *Terebralia palustris* lives among mangroves in shallow water.

FAMILY: CERITHIIDAE: Ceriths or Horn Shells

Ceriths are important members of the shell fauna on shallow Red Sea reefs where they live in a wide range of habitats, feeding on detritus and decayed algae. Their long spirally coiled shells have many whorls and the surface is usually well-sculptured. A distinctive feature of the aperture is the recurved siphonal canal. There is considerable variation, even within the same species and this renders positive identification a difficult task.

Rhinoclavis kochi has a slender shell with weakly developed axial costae. It lives in mud and fine sand in shallow water, particularly on sea-grass beds where it may be quite abundant.

Rhinoclavis fasciatus occurs in coarse sand in lagoons and in similar sediments mixed with coral debris in the fore reef zone.

Cerithium ruppelli has a roughly sculptured shell. It lives in mud and fine silt in shallow water, often inside harbours.

C. erythraeonense is another shallow water species occurring in sea-grass beds and sandy lagoons.

FAMILY: TRIPHORIDAE

Shells from this family have the unique characteristic among marine gastropods of being sinistrally coiled, so that the aperture is left of the axis instead of being to the right. Like Ceriths, their shells are in the form of a spire. They live in close association with sponges on which they apparently feed. Although they can be quite abundant they are frequently overlooked as a result of their small size and frequently concealed habitat.

More than twenty species occur in the Red Sea and they include members of the genera: *Triphora, Mastonia, Cantotriphora, Viriola* and *Iniforis*.

Triphora tricincta is a relatively common species occurring under micro-atoll coral heads and in crevices on reef platforms.

FAMILY: EPITONIIDAE: Wentletraps

The distinctive sculpturing on the shells of wentletraps renders them unmistakable. They have a series of thin ridges which appear like a spiral staircase. The shells are usually pale or white with a dark horny operculum. They feed on live corals or anemones and possess a cuticularised oesophageal lining which possibly prevents injury from nematocysts.

Epitonium scalare is one of the larger species (shell approx. 5cms). Its shell has conspicuous axial ribs and a round aperture. When disturbed it secretes a purple fluid.

Epitonium lamellosum is relatively common on sea-grass beds in the central Red Sea.

Epitonium aculeatum has diagonal, even ridges.

Bedouin fishing hut whose wall is made of *Strombus* shells.

Close-up of fishing hut wall.

Hipponix conicus

FAMILY: JANTHINIDAE

Janthina janthina: Common Purple Sea Snail.
Shell is approximately 2cms long and 3.8cms wide, globular and fragile. It is pale violet above and deep purple below, and there are three to four sloping whorls with a large aperture. *Janthina* floats at the surface of the sea with the aid of a raft of mucus bubbles which it secretes and from which it hangs. It feeds on velellid jellyfish.

FAMILY: STROMBIDAE: Conch Shells

Juvenile conch shells resemble cones and in adults a cone-shaped, frequently sculptured spine persists. The aperture is elongated and the outer lip has a distinct notch through which, in living forms, an eyestalk may protrude. They are herbivorous molluscs and frequently occur among sea-grasses, where they may sometimes be observed to "hop" along the sea-bed using their sickle shaped, horny operculum like a pole-vault. This action may also be observed by turning a stromb upside down (i.e. aperture uppermost) and watching how it rights itself.

Strombus tricornis is endemic to the Red Sea and Gulf of Aden. It is a heavy shell (up to 15cms) which is common on shallow sandy areas, whether these are associated with sea-grasses or with coral rubble.

Strombus gibberulus albus occurs on shallow sea-grass beds where the sediment is quite fine. It may be locally abundant and is preyed on by octopus.

Strombus fasciatus is a relatively small stromb (3-5cms) with a triangular body whorl, the shoulder of which carries conspicuous nodules. It is white with fine dark brown spiral bands and spots. It has a similar habitat to the previous species.

Lambis truncata sebae: Spider Conch or Finger Conch
One of the largest conch shells with fully grown specimens reaching 40cms in length. Such adults are quite unmistakable but young juveniles could be confused with other families.

Its preferred natural habitat is probably the shallow reef-flat but it is often intensively collected in this most accessible of reef locations. It also occurs on coarse gravel or among coral rubble in shallow lagoons, and on similar substrata in the fore-reef.

Tibia insulaechorab: "The Arabian Tibia"
It has a long, smooth, spirally coiled, ebony coloured shell which has an outer lip bearing a number of short, finger-like processes. It is normally a very elusive species, but in spring divers have reported finding it on shallow silty bottoms of sheltered inlets. It seems to migrate to these areas in order to lay its eggs. These are ejected in a delicate, flattened, spiral egg-string which is frequently deposited on the summit of the volcanic-shaped burrows of an unidentified species (possibly a hemichordate). Regular eruptions of sand from the apex of the mounds ensure that soon after egg-laying is complete, the egg-string is covered by a protective thin layer of sand.

Terebellum terebellum is the only stromb which lacks a stromboid notch. It has a smooth, variably patterned, torpedo-shaped shell (up to 7cms in length). It lives buried in coarse lagoon sand, through which it is able to propel itself with remarkable rapidity. It is even able to swim for short distances.

Strombus mutabilis and *S. erythrinus* also occur in the Red Sea, and are among those illustrated.

FAMILY: HIPPONICIDAE: Hoof Shells

Cap-shaped shells with a backward pointing apex and with radial ribs. They are found attached to stones, corals and to other shells.

Hipponix conicus lives as an ectoparasite on gastropods and on spines of the sea-urchin *Phyllacanthus imperialis*. It is a relatively rare shell which is most often found near the reef-crest.

STROMBIDAE: CONCH SHELLS

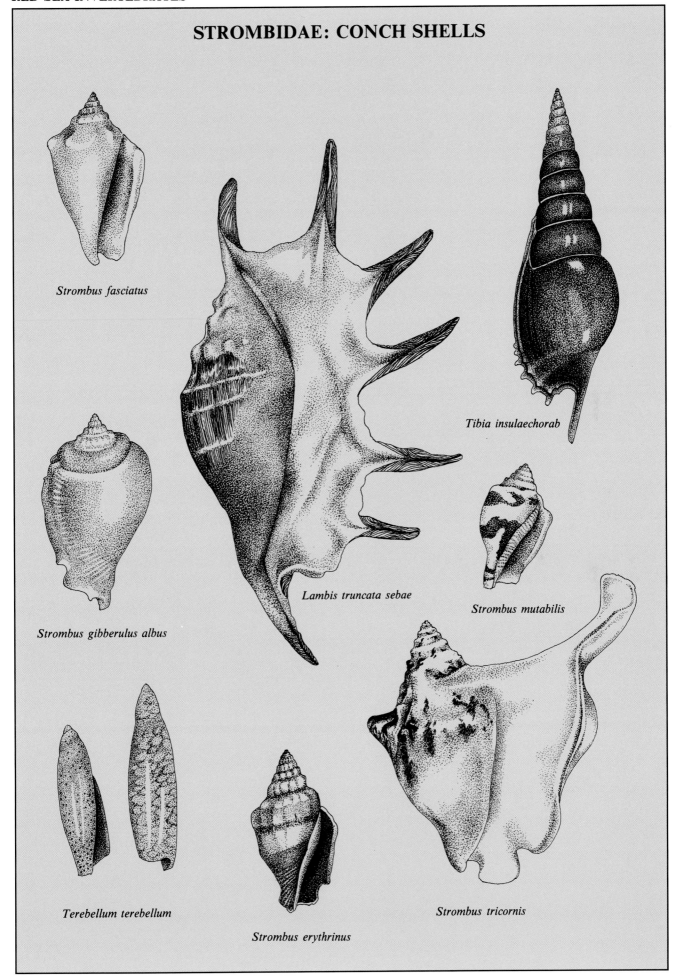

Strombus fasciatus

Tibia insulaechorab

Strombus gibberulus albus

Lambis truncata sebae

Strombus mutabilis

Terebellum terebellum

Strombus erythrinus

Strombus tricornis

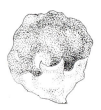

Cheilia equestris

Cypraea nebrites

Cypraea nucleus

Arabian cowrie: *Cypraea arabica*

FAMILY: CALYPTRAEIDAE: Cup and Saucer Shells

Similar to limpets but with a cup-shaped structure inside.

Cheilia cicatricosa has a shell which measures approximately 3.5cms and is sculptured with ridges.

Cheilia tortilis occurs beneath coral heads, in crevices and on the cemented reef platform.

FAMILY: CYPRAEIDAE: Cowries

Cowrie shells have attracted and fascinated Man since earliest time. They are probably the most beautifully patterned and glossiest shells on the reefs and there is a wide range of species, each with its own characteristic shell pattern. They include herbivorous, omnivorous and carnivorous forms and most of them remain hidden during daytime emerging at night. The only similar family is the *Ovulidae* and they can be separated from *Cypraeidae* by virtue of the fact that both lips of the elongate aperture are toothed in cowries, whereas only the outer one carries teeth in *Ovulids*.

As juveniles, cowries do resemble some other gastropods but as the shell grows the last whorl envelops the younger shell and forms the typical egg-shaped cowrie shell.

When searching for cowries it is important to remember that the distinctive pattern is frequently obscured by a thin layer of mantle which the mollusc spreads over the shell. It is this which keeps the shell glossy and free from epifauna.

Cypraea grayana: Arabian Cowrie

This has a creamy-white shell with large, darker blotches. The upper surface of the shell has a pattern of dark brown longitudinal lines interrupted by reticulations. There is some doubt whether this species is separate from *Cypraea arabica* which has a pattern of lines without the spots. Mean shell size is 5cms.

Cypraea carneola is a relatively large cowrie, reaching in excess of 9cms. It is light brown in ground colour with creamy white transverse bands. It occurs frequently on the underside of littoral rocks and boulders.

Panther cowrie: *Cypraea pantherina*

Cypraea nebrites has a mottled pattern of light spots on a brown background, together with two prominent dark blotches which are visible from above but do not extend onto the base. It is relatively common intertidal species.

Cypraea nucleus has a shell (approx. 2cms) which superficially resembles *Trivia* with prolonged teeth covering the ventral surface and continuing dorsally as irregular transverse ribs.

Monetaria moneta is a small cowrie generally about 2cms long. Periphery is somewhat flattened so that the centre appears to bulge. Creamy white to yellowish. Found under boulders and in sea-grass beds, but is relatively scarce on central Red Sea reefs despite its frequent occurrence in sub-fossil reefs in the same region. It is reported to be common on reefs in the Massawa region.

Monetaria moneta

Monetaria annulus: Ring Cowrie
Similar to *Monetaria moneta* but differentiated by the orange or yellow ring which surrounds the central bulge. Pale blue-grey inside the line and cream outside. It occurs under coral heads and in coral crevices in shallow water.

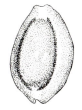

Monetaria annulus

Cypraea pantherina: Panther Cowrie
This species is endemic to the Red Sea and is closely related to the Indo-Pacific Tiger Cowrie, *Cypraea tigris*. It differs from this well-known species by being smaller and narrower with more vertical sides and more numerous teeth. It is white to greyish or brownish with many small dark spots. The base is white.

Cypraea isabella has a distinctive shell pattern. It is somewhat cylindrical in shape and fawn grey with long, dark, interrupted lines. The tips are orange and the shell base is white. In life the mantle is jet black.

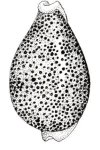

Cypraea pantherina

Cypraea isabella

Cypraea fimbriata is a whitish cowrie with a pattern of mottled specks and banding in brown and grey.

Cypraea turdus: Thrush Cowrie
A common Red Sea cowrie which is well camouflaged by its mantle. It is cream coloured with many light brown dots and the base is white. Mean shell size is 2.8cms.

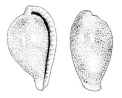

Pseudosimnia marginata

Polinices melanostomus

Triton's Trumpet Shell: *Charonia tritonis*

FAMILY: OVULIDAE: Egg Cowries

These are cowrie-shaped shells which can be distinguished by virtue of the fact that only the outer lip carries teeth. They are usually found in association with soft corals or sponges.

Calpurnus verrucosus has a shiny white shell (approx. 4cms) with pinkish tubercles at each end and coarse teeth along the outer labial lip. It is found on the fore-reef terrace, usually in association with soft corals such as *Sarcophyton*.

Other Red Sea ovulids include the smaller species: *Pseudosimnia marginata* and *Pseudocypraea adamsonii*.

FAMILY: NATICIDAE: Moon Snails

Moon shells are predators of other molluscs, which they locate by ploughing through the sand with the aid of a large foot. Utilising a well developed radula, they bore a neat round hole through the shell of their prey and then proceed to consume their victim via this narrow opening.

Natica onca. Shell (approx. 3cms) has a large end whorl and an umbilicus partly filled by a callus. It is shiny whitish with spiral rows of red or brown patches; the semicircular operculum shows strong grooves parallel to the outer lip. Found burrowing in the fine sand of shallow lagoons.

Polinices melanostomus: Black Mouthed Moon Snail
Shell (approx. 5cms) dark brown with reddish-brown horny operculum. It is locally common on sea-grass beds.

HORST MOOSLEITNER

FAMILY: CASSIDIDAE: Helmet Shells / Bonnet Shells

These are relatively heavy and sometimes large shells which generally remain hidden during the daytime but emerge at night to hunt their prey of sea-urchins. The shell aperture is relatively narrow and the outer lip is occasionally toothed. The operculum may be fan-shaped. The siphonal canal is directed towards the posterior.

Phalium bisulcatum is a whitish shell (2-6cms) with five to six rows of red square spots.

Casmaria ponderosa has a characteristic pattern of brown spots on the thickened callus. The lip is sharply toothed. It occurs on fine silt or mud.

Cassis cornuta. Horned Helmet is a large, heavy, helmet shell (35cms) which occurs from shallow to deep water on sandy areas.

Casmaria ponderosa

FAMILY: CYMATIIDAE: Triton Shells

Large shells whose widespread distribution throughout the Indo-Pacific region results from their unusually long larval life. Their veliger larva remains in the plankton for approximately three months. They are carnivorous species and feed on sea-urchins, sea-stars, worms and other invertebrates.

Charonia tritonis: Triton's Trumpet Shell
It is one of the largest Red Sea molluscs and has unfortunately attracted shell collectors for many years. The shiny, patterned body whorl is gracefully curved and the spire is long and tapered, generally with the end broken. The inside opening is bright orange and there are sharp teeth on the lip.

This species is one of the few proven predators of the Crown of Thorns Sea-star: *Acanthaster planci*. It has been suggested that over-collecting of it on Indo-Pacific reefs provided a trigger for population explosions of the coral predating starfish. Whether or not this is the case, there is no doubt that triton shells should be a protected species. It is found in shallow to moderate depths, associated with corals and coral rubble.

Distorsio anus has an extremely fragile shell with a hairy periostracum. The ventral side forms a flattened shield and the aperture is narrowed by folds and teeth; colour whitish with brown bands and spots. It is found under shallow boulders and in crevices on the reef platform.

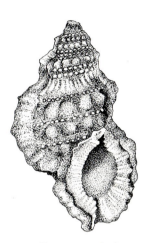

Bursa granularis

FAMILY: BURSIDAE: Frog Shells

These resemble Triton shells but they have a second canal on the opposite side of the aperture to the usual siphonal canal. The outer lip is frequently toothed and the columella folded. They are shallow water carnivorous species which feed, in particular, on worms.

Bursa granularis has a shell (approx. 5cms) which is heavily and irregularly sculptured with spiral ridges and tubercles. The siphonal canal is short and the operculum is pale coloured. It is a relatively common species occurring in shallow water under boulders and micro-atolls.

Tutufo bubo is the largest member of the family in the Red Sea. The shell is quite variable but the secondary canal is not as distinct as in *Bursa granularis*. It lives in shallow to moderate depths in association with coral and rubble.

Tonna perdix

Partridge Tun: *Tonna perdix*

FAMILY: TONNIDAE: Tun Shells

Carnivorous molluscs, usually confined to deep water, Tun shells have large thin shells and no operculum. They live on sandy bottoms and feed on sea-cucumbers and various other invertebrates.

Tonna perdix: Partridge Tun
The shell of this large species is very fragile. It lives on sand of lagoons or in the fore-reef zone where it remains buried during daytime but emerges at night to hunt for its prey of crustaceans, sea-cucumbers or gastropods which it attacks with a highly acidic saliva.

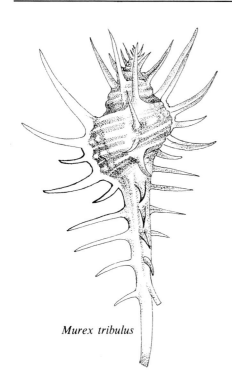

Murex tribulus

Pyrene testudinaria

Engina mendicaria

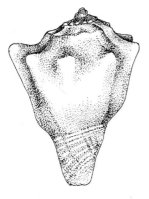

Volema pyrum nodosum

FAMILY: FICIDAE: Fig Shells

Not unlike Tun shells to which they are quite closely related, these have a thin but strong shell with a wide open aperture which tapers into a siphonal canal. The surface is covered with spiral ribs and axial cords which form a net-like pattern. *Ficus subintermedius* has been recorded from the Red Sea but is considered rare.

Order: Neogastropoda

These marine prosobranchs include a number of familiar families including whelks, murex shells, olives and cone shells. They are spirally coiled gastropods which usually have a long siphonal canal. In most cases the proboscis is very well developed and the radula may be adapted for injection of poison into their prey.

Suborder: Stenoglossa

These have a radula with three or less large teeth in each row.

FAMILY: MURICIDAE: Murexes: Comb Shells

These are usually characterized by spiny projections, the development of which is dependent upon the species and the environmental conditions affecting a particular specimen. Thus, those individuals of a species which occur in a sheltered habitat may have very long, pointed spines, whereas the same species from a less protected habitat may have short, blunt spines. They are carnivores which attack their molluscan prey by drilling holes through their shells and releasing a purple substance ("purpurin") which causes paralysis.

Murex tribulus has a variable shell (approx. 10cms) with a long siphonal canal and long, pointed spines. They occur mainly in deep water (e.g. 30m to 250m) but migrate into shallower water to lay their eggs.

Chicoreus ramosus is a shallow water species with short, blunt spines projecting from its robust shell. It has a light ground colour with brown patches or spiral lines. The siphonal canal is pink.

FAMILY: PYRENIDAE: Dove Shells

These are tiny shells (usually less than 2.5cms) which have shiny, attractively patterned shells. The aperture is usually elongate and the outer lip thickened and toothed.

Pyrene testudinaria has a smooth, shiny shell with a highly variable pattern of brown patches on a white background. It is a herbivorous species which feeds at night and may be found during daytime, concealed under boulders, or in crevices, from the reef platform down to around 20m depth.

FAMILY: BUCCINIDAE: Whelks

Whelks are primarily carnivorous scavengers which feed on bivalves, worms and carrion. They have strongly built, robust shells and usually possess a large horny operculum.

Engina mendicaria is a small whelk (2cms) which has a thick shell with the final whorl inflated. The surface is smooth with weak axial ribs and shoulder knots; the siphonal canal is short; colour is white to yellowish with black spiral bands and the upper whorls plain white. It occurs in the low intertidal and subtidal levels, on beach rocks, under boulders and in crevices on the inner reef-flat.

Cantharus fumosus is a locally common species which is found under littoral rocks or boulders, and in crevices or under coral heads.

FAMILY: MELONGENIDAE: Crown Conchs

Volema pyrum nodosum is a carnivorous mollusc which feeds on other molluscs and worms in shallow, sandy or muddy areas. It is frequently associated with mangroves or among sea-grasses. The adult shell measures approximately 5cms.

FAMILY: NASSARIIDAE: Mud Snails

Mud snails have small, robust shells which may be quite heavily sculptured. The aperture is relatively wide, with the outer lip toothed along the inside and the operculum is correspondingly serrated. They have a very well developed olfactory sense which enables them to locate dead organisms from as far away as 30m. Taxonomy of the group is complicated by the degree of environmentally-induced variability which exists within each species.

Nassarius gemmulatus has a broad oval whitish shell (approx. 3cms) with strong axial ribs and spiral grooves. It lives on sand.

Nassarius arcularis plicatus: Little Box Dog Whelk
Olive to white with strong pleats on the spire. It lives on mud or fine silt and is often found in association with sea-grasses.

Nassarius arcularis plicatus

FAMILY: FASCIOLARIIDAE: Spindle or Tulip Shells

Members of this family have an aperture which continues into a long siphonal canal and shells which are usually ornamented with spiral ridges and axial ribs. The periostracum is frequently brown and flaky or fibrous in texture. They are carnivorous, feeding on molluscs, worms and other small invertebrates. Some species are relatively large (e.g. 14cms), while others are only two or three centimetres in length.

Pleuroploca trapezium is a relatively large tulip shell (14cms) with a wavy lip. They live in sand and among coral rubble on the inner reef-flat.

Latirus polygonus has a yellow-brown shell with spiral cords and nodules. There are black patches on vertical ridges. It occurs on fossil beachrock and occasionally on structures such as reef markers, jetties, etc.

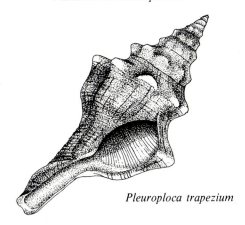

Pleuroploca trapezium

Tulip shell: *Pleuroploca* sp.

Oliva bulbosa

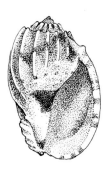

Harpa amouretta

Mitra fasciolaris

Xenoturris cingulifera erythraea

Vexillum coronatum

FAMILY: OLIVIDAE: Olive Shells

Smooth, polished, cylindrical shells with variable markings. They are active carnivores which remain hidden under fine sediment during the day, but emerge to hunt for their food at night.

Oliva bulbosa has a shell with an extremely variable colour pattern. A complete series would range from mainly white, to brown, and even black markings.

Other species include *Oliva elegans; Ancilla lineolata; A. acuminata* and *A. cinnamocea.*

FAMILY: TURBINELLIDAE (VASIDAE): Vase Shells

The shell is solid and spirally shaped with a narrow aperture. The columella usually has several plaits or folds and the whorls are ornamented with nodules.

Vasum turbinellus occurs in sand or silt on the shallow lagoon floor, among sea-grasses or associated with coral rubble. Its shell is quite often heavily encrusted.

FAMILY: HARPIDAE: Harp Shells

These have attractive shells in which the body whorl is bulbous and sculptured on the outside, with a series of neat longitudinal ridges somewhat reminiscent of harp strings. The aperture is wide, and the large, muscular foot forms an effective burrowing organ. They are carnivores which feed mainly on small crustaceans, which they engulf in a layer of saliva and sand. If attacked, they are able to cast off the rear end of their foot, which presumably acts as a diversion while the harp shell makes good its escape.

Harpa amouretta occurs sporadically in a wide range of habitats from the shallows to relatively deep water.

FAMILY: MARGINELLIDAE: Margin Shells

These small, polished shells are named after their thickened lip. They are predatory forms which lack an operculum.

Marginella monilis occurs in shallow water buried in sand or under stones.

FAMILY: MITRIDAE: Mitre Shells

Mitre shells are carnivorous gastropods whose tracks are often visible on sand. They plough their way through the sediment in search of their prey of small invertebrates or of organic detritus on which some species scavenge. The inner lip of the aperture (part of the columella) has several folds. Their common name is derived from the resemblance of their elongated shells to the shape of a Bishop's mitre.

Subanciella annulata has a shell (2 to 3.5cms) with typical strong and narrow spiral ribs and double rows of small depressions in between. There are four to six columellar folds. It occurs on coarse sand from shallow to deep water.

Mitra cucumeria has a shell (approx. 3cms) with a short spire, strong axial ribs and three to four columellar folds. It is dark brown with indistinct lighter bands and irregular small patches of white.

Mitra fasciolaris is a Red Sea endemic species which occurs intertidally among coral rubble and coarse sand.

Mitra litterata: The Lettered Mitre
It is a relatively common mitre which occurs intertidally and in shallow water under rocks, in crevices and among algae.

FAMILY: COSTELLARIIDAE

Vexillum leucozonias occurs intertidally in the shallows and down to around 20m depth on sand and coralline algae.
Vexillum coronatum is found on white sand at moderate depths.

Suborder: Toxoglossa

This suborder of the Neogastropoda includes those forms, such as cone-shells, which possess a radula with two rows of loosely arranged harpoon-like teeth which are adapted for paralysing their prey by the injection of poison.

FAMILY: TURRIDAE: Turrids

These are small, spindle-shaped shells which superficially resemble Fasciolariidae, but they have venom glands and are more closely related to cone-shells. An identifying factor is the presence of a slit in the upper part of the outer lip (the "turrid notch"). They are carnivorous species with representatives occurring from the intertidal to deep water.

Drilla flavidula. It has a high shell (4-6cms) with a posteriorly directed siphonal canal and a reticulated pattern formed by prominent axial ribs and numerous spiral cords. Colour creamy yellow with indistinct brown bands. Occurs in silt and fine sand at moderate depths.

Tritonoturris cumingii. It is a fragile turrid which is found in sand pockets, under coral, at around 20m.

Xenoturris cingulifera erythraea is a Red Sea subspecies.

FAMILY: CONIDAE: Cone Shells

Cone shells display quite a wide range of colour pattern variations within a single species and this has given rise to taxonomic difficulties and, in many cases, several different names being given to what is later proven to be one species. The general cone-shape is a universal characteristic within the family but environmental conditions may influence its precise size and shape and thus cause intra-specific variations of form as well as colour. A thin, skin-like periostracum often obscures the shell's underlying pattern. Some cone-shells possess a venomous sting which is powerful enough to kill a person. They normally use this venom apparatus to paralyse their prey of worms, molluscs or fish.

Conus pennaceus has a pattern of white triangles on a black shell. The density of the pattern varies considerably and has caused much confusion within the species. It has a broad range of habitats from shallow to deep water where it lives on sand. It feeds on other molluscs. Up to six radula teeth may be injected into the same prey organism.

Conus geographus is considered to be one of the most potentially dangerous cone shells and has been the cause of several documented human fatalities. It has a relatively thin shell bearing an intricate pattern of creamy, brownish or red patches, stripes and spots on a white background. It is a fish-eating cone shell which occurs among corals, especially in the upper fore-reef.

Conus vexillum is a relatively large (e.g. 15cms), heavy cone shell with broad shoulders and a predominantly streaky orange-brown colour with two bands of irregular white patches and a white and brown patterned spire. It occurs in both shallow and deep habitats.

Conus virgo has a dull cream or yellow shell which fades into purple towards the base. In life it is normally covered by a thick brown periostracum which obscures the true shell colour. It may grow to 15cms or more and is found on sand in shallow water, frequently associated with sea-grasses.

Conus fulgetrum has a dark violet inner shell at the aperture and a pattern formed by zig-zag stripes. It occurs among the sediment of sea-grass beds and in sand under coral boulders.

Conus taeniatus is a relatively common shallow water cone shell which occurs in sand among the roots of sea-grasses, as well as under rocks in the intertidal and shallow subtidal. Its colour pattern is formed by a spiral array of dashes and one, or several, indistinct darker bands on a white background.

Conus flavidus is another species of cone shell present in the Red Sea.

Conus pennaceus

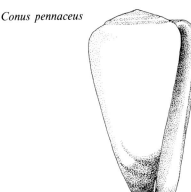

Conus virgo

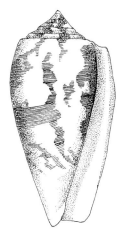

Conus striatus

Conus flavidus

Sand cone: *Conus arenatus*

Textile cone: *Conus textile*

Conus textile has a white shell (4-12.5cms), with a pattern of angular brown patches. Lives in a wide range of shallow water reef habitats including on fine sediment in protected lagoons, among sea-grasses, on coral rubble and partly covered by sand under boulders in the shallows. It is considered dangerous to Man. During daytime, it remains on, or partly buried in sand under rocks, but crawls actively over the sand and coral at night, when it feeds.

Conus arenatus: The Sand Cone has a thick shell (5-7cms) with small black spots on a white background together with two indistinct spiral bands. It occurs on sand in lagoons and in shallow or intertidal areas.

Conus striatus has a more curved, cone shaped shell which is patterned all over by deep prominent striae. It is whitish, stained with rose and irregularly streaked and mottled with black. It occurs in sand, in sheltered bays, generally at shallow depths, sometimes under rocks. It has an efficient "harpoon" and venom apparatus and is considered as potentially dangerous to Man.

FAMILY: TEREBRIDAE: Auger Shells

Long slender shells with a pointed apex and a small aperture with a simple sharp lip and a single fold on the columella. Many species are beautifully patterned and since they lack a periostracum the shell colours are perfectly displayed on live individuals. They are carnivorous, feeding on worms and other small invertebrates which they hunt in the sand. Some species have a cone-shell like venom apparatus.

Terebra maculata is the largest of the Red Sea augers, growing to at least 28cms in length. It has a creamy white shell with two spiral bands of dark brown spots (or flame-shaped patches) on each whorl. It lives on sand in shallow regions (e.g. 7m) especially in lagoons of offshore reefs, and on sea-grass beds.

Terebra affinis has a much smaller, narrow shell (approx. 6.5cms) with a relatively smooth surface and brown pattern. It occurs in coarse lagoon sand in shallow water.

Terebra affinis

Order: Bullomorpha

Not all members of this order possess a shell. In some cases it is internal and in others it is completely absent. Most species are burrowers and are particularly adapted for this mode of existence with a flattened head and large foot.

FAMILY: ACTEONIDAE: Pupa Shells

These small, solid shells have large body whorls, short spires and elongated apertures. The mollusc can withdraw fully into its shell.

Pupa solidula has a strongly ridged shell (approx. 3.5cms) with reddish spots (sometimes ranging from fawn to black). It has an operculum. Lives in sandy areas, especially in lagoons and back reef areas.

Pupa solidula

FAMILY: BULLIDAE: Bubble Shells

Generally small molluscs with smooth shells, whose bubble-like shape is created by an enlarged body whorl and depressed spire. The aperture is wide and in life mantle lobes protrude and partially cover the shell. They are herbivorous and occur in calm, sheltered waters, often among sea-grasses.

Bulla ampulla burrows in sand in shallow water, especially among sea-grasses. Shell small (e.g. 1.8cms long).

FAMILY: HYDATINIDAE: Rose Bubble-shells

Thin, globular shells into which the animal is able to withdraw. The spire is depressed and the final body whorl inflated to give a more or less spherical shape. Living forms cover part of the shell with fleshy pink extensions of the mantle and foot which have been likened to the petals of a rose.

Hydatina physis may be found in sea-grass beds where it preys upon polychaetes.

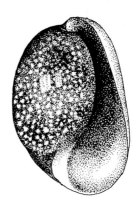

Bulla ampulla

FAMILY: ATYIDAE

Fragile, egg-like shells with sunken spires. In life shell partially concealed by cephalic parapodial lobes. They usually burrow in sand.

Atys cylindricus is a very delicate white or light fawn coloured, translucent shell (approx. 3cms).

FAMILY: SCAPHANDRIDAE: Scaphandrids

They are small elongate molluscs, e.g. *Cylichna girardi*.

FAMILY: AGLAJIDAE: Aglajids

Members of this family have a small shell enclosed by parapodia near the posterior of the animal. The general appearance is like that of a slug but there are paired asymmetrical projections at the rear. They are carnivores and live in shallow, sandy areas, e.g. *Aglajia cyanea*.

Atys cylindricus

Pyramidella sulcata

Aplysia sp. laying yellow egg strings

Sea hare: *Aplysia* cf. *dactylomela*

FAMILY: PHILINIDAE: Philinids
Thin colourless, translucent ear-shaped shell with wide aperture. The animal cannot withdraw. *Philine vaillanti* is found in the Red Sea.

Order: Pyramidellomorpha

FAMILY: PYRAMIDELLIDAE: Pyramidellids
These have solid, spiral shells with long spires. They are carnivorous species which prey on various bottom living invertebrates or else they live as ectoparasites on molluscs, sponges or worms.

Pyramidella sulcata occurs among coral rubble or in shallow crevices under coral heads.

Order: Thecosomata

FAMILY: CAVOLINIIDAE
Pelagic molluscs which may be abundant at certain times of year. Their shells, rather than spirally coiled are bilaterally symmetrical and funnel-shaped. The foot is extended into two large parapodial lobes which are used for swimming.

Cavolinia longirostris occurs throughout the Red Sea and its shells are often washed up on the shore-line.

Order: Aplysiomorpha

FAMILY: APLYSIIDAE: Sea Hares
Large, slug-like animals with four tentacle-like structures on a prominent head. Shell internal or absent. Foot used for creeping over the substrate and for swimming. Sea hares are herbivores which feed on algae. When disturbed they emit a purple fluid. Egg masses are laid in long gelatinous strings.

Dolabella auricularis is probably the most frequently encountered large sea hare, which lives on sea-grass beds in shallow water.

Figure 31: *Notarchus indicus:* method of propulsion.

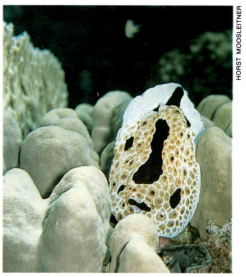

Pleurobranchus sp. in daytime

Pleurobranchus sp. expanded at night

Notarchus indicus can be extremely abundant (up to 50/m²) on *Halophila* in sheltered shallow water (1-4m), especially from February to mid-April with a peak occuring (in the Gulf of Aqaba) during the first half of March. It has a typical sea hare form and is generally 3-4cms long. It feeds upon the layer of detritus which generally covers this alga in sheltered regions.

It has a unique form of swimming which, instead of flapping the parapodia or bending the body up and down, involves a form of jet-propulsion. The parapodia, which are folded back over the body, are joined dorsally to form a sac which is open anteriorly. When it is going to swim it folds the foot, retracts its head, and inflates the dorsal sac with water so that it resembles a ball. This causes its centre of gravity to shift and it rolls forward bringing the opening of the parapodial sac to a downward pointing position. Expulsion of water from the inflated sac then propels the sea-hare up into the water column. As it does so it somersaults until the next intake and subsequent expulsion, which propels it again through a series of somersaults (see figure 31).

Stylocheilus longicaudus is a smaller aplysiid. Body is slender with a distinct, long pointed tail. The colour pattern consists of longitudinal dark striations and red-blue spots.

Order: Pleurobranchomorpha

FAMILY: PLEUROBRANCHIDAE: Side-gilled Slugs
These are soft bodied molluscs which usually (but not always) lack a shell. They have a long feathery gill on the right hand side, between the foot and the mantle. Many species can secrete a strong acid and have calcareous spicules in the skin.

Berthellina citrina occurs under stones in shallow water.

Order: Sacoglossa

FAMILY: ELYSIIDAE: Elysiids
Slug-like with a pair of tentacles on the head and a flattened body with no shell. Parapodial lobes folded to cover the back.

Elysia decorata is probably a Red Sea endemic species. It occurs under coral rocks or in crevices in shallow water.

Elysia latipes is also present.

Order: Nudibranchia: Nudibranchs

These Opisthobranch molluscs have no shell but may have spicules embedded in the skin. They have bilaterally symmetrical slug-like bodies which frequently carry finger-like extensions known as cerata. Nudibranchs presented a problem for early biologists since, in the absence of a shell, their descriptions had to be based upon soft bodies whose colours usually fade when they are preserved. Thus the specimens upon which many species were described were shrivelled lumps of flesh which bore little resemblance to the beautiful live creatures which ardent field-workers had collected with such enthusiasm and fascination, often under quite difficult conditions. In a number of instances however, biologists actually painted the collected specimens when they were still alive or else took detailed notes upon their natural colours before preserving them.

This situation has been altered in recent years by the advent of underwater photography. The true magnificence of nudibranchs is emphasized by the colour slides published in this book. Strangely, however molluscan taxonomists can still experience difficulties with making positive identifications because the characteristics upon which most species have been described are based upon internal anatomy rather than external appearance.

Nudibranchs are all predators and they feed upon a wide range of invertebrates but many species show quite distinct dietary preferences, resulting in them regularly occurring in association with certain species. Their often bright colours may have protective significance in serving to warn predators of their unpleasant characteristics, derived from skin gland secretions of a distasteful fluid or of stinging cells which some coelenterate-feeding nudibranchs divert from their gut to their cerata.

A generalized nudibranch is illustrated in figure 32.

Suborder: Dendronotacea

Members of this sub-order have lateral gills and tentacles on the head which can be retracted into special sheaths. The mouth is usually surrounded by a veil of skin.

FAMILY: TETHYDIDAE

Melibe bucephala. Adults which occur in springtime, measure 10-12cms. Full size is attained in about three weeks from the juvenile stage (measuring about 1cm). It can easily be mistaken for an algal cluster and is more or less transparent with branched appendages on the dorsal side. It may be locally abundant (e.g. 20/m²) in sheltered harbours where the alga *Hydroclathrus* grows. It has an oral hood which is an extension of the head. This has two rows of tentacles along its margin. There is no radula but two chitinous mandibles. It lives on thin filamentous algal filaments such as *Sphacelaria*, *Enteromorpha* or *Hydroclathrus* growing in calm water. They close their oral hood around algal filaments and use their tentacles to wipe the algal threads clean of detritus. When disturbed it can swim by bending its body laterally.

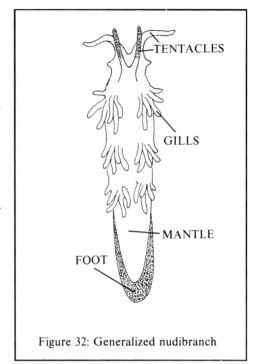

Figure 32: Generalized nudibranch

Opposite: Spanish Dancer: *Hexabranchus sanguineus;* left: Nudibranch: *Halgerda willeyi;* above: Nudibranch: *Chromodoris quadricolor* on black sponge.

F. JACK JACKSON

Nudibranch: Spanish Dancer: *Hexabranchus sanguineus*

Eggs of Spanish Dancer

F. JACK JACKSON

Suborder: Doridacea
   This suborder comprises the largest group of nudibranchs. They are dorsoventrally flattened and one pair of tentacles can be withdrawn, but tentacular sheaths are not usually present. The gills are normally arranged in a circle, posteriorly, around the anus. In more advanced forms gills can be withdrawn into a special cavity.

FAMILY: HEXABRANCHIDAE: Spanish Dancers
Hexabranchus sanguineus is commonly known as the Spanish Dancer since it is coloured bright red with a white fringe and is a most graceful swimmer. It is relatively large, up to 15cms long. It feeds upon sponges and ascidians and is often accompanied by a commensal shrimp: *Periclimenes imperator*.

FAMILY: GYMNODORIDAE: Gymnodorids
   Slug-like, no shell, soft bodies, e.g. *Gymnodoris impudica*.

FAMILY: POLYCERIDAE: Polycerids
   Differentiated from Gymnodorids by internal differences including structure of radula teeth. Mainly feeds on bryozoans, e.g. *Nembrotha limaciformis*.

FAMILY: PHYLLIDIIDAE: Phyllids
Slug-like nudibranchs with numerous gills. Protuberances on mantle; body tough and somewhat rigid.

Phyllidia varicosa approximately 5cms, with three longitudinal ridges on dorsal surface separated by black grooves. Towards edge of mantle white ridges run transversely. Ridges carry rows of bright orange tubercles. Rhinophores orange. Ventrally, mantle and foot are grey with darker line along middle. Grooved oral tentacles tinged with yellow at tips.

Phyllidia pustulosa is about 4cms long; dense black ground colour with white tentacles surrounded by a white halo. Median row of tubercles with elongated halos along keel of the back; halos of last two tubercles confluent.

FAMILY: DENDRODORIDIDAE: Dendrodorids
Dendrodoris rubra is approx 2-5cms long. Body oval, smooth, mantle margin thin and crenulate. Red with mantle darker than gills and foot. Larger forms with grey to blackish blotches. Rhinophores red with white tips.

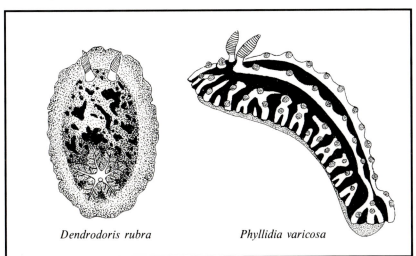

Dendrodoris rubra   *Phyllidia varicosa*

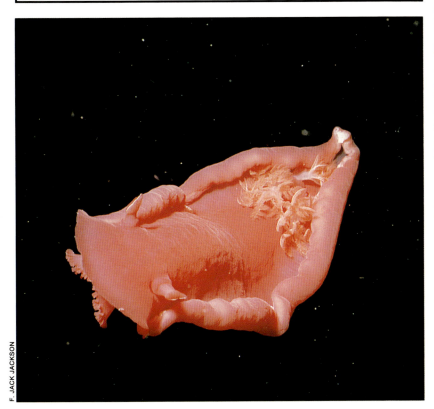

Phyllidia bourguini

Phyllidia varicosa

Phyllidia cf. *elegans*

Above: Doridacean nudibranch
Left: Spanish Dancer at night

Top and bottom left: *Chromodoris* sp.; top and bottom right: *Chromodoris quadricolor*; opposite: *Casella cincta*.

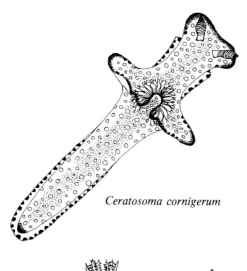

Ceratosoma cornigerum

Hypselodoris infucata

FAMILY: CHROMODORIDIDAE: Chromodorids

Hypselodoris infucata measures up to 4cms long. Edge of mantle clear dorsally with irregularly arranged Prussian blue spots and several larger yellow spots. More centrally the mantle is suffused with Prussian blue with dark blue spots and a few yellow spots. Centre of back is white with about twenty yellow spots and a few blue spots. Rhinophores white or mauvish at base and bright orange at ends. Sides of body blue or bluish grey, tail whitish grey with a tint of blue dorsally. Yellow and blue spots on tail with yellow ones surrounded by ring of white. Foot rounded and transversely grooved anteriorly; tinged with bluish grey but front edge purplish blue. Oral tentacles clear grey with tips which are sometimes orange.

Chromodoris annulata is 2cms to 3.5cms long. Body colour white, with yellow spots all over dorsal surface of mantle, as well as side of foot and dorsal side of posterior end of foot. Mantle edged with dark purple and similar coloured rings round gills and rhinophores. Anal papilla white with purple ring at orifice. Foot white and ventral surface of mantle suffused with pale purple.

Ceratosoma cornigerum has two anterior lobes in front of rhinophores and two lateral lobes behind gills. Dorsal hump behind gills. Tail long and narrow, roughly equivalent in length to that of remainder of body. Ground colour cream with brownish-red mottling on dorsal surface. Mantle edged with purple and sides of foot have purple spots. Orange spots scattered over body. Oral tentacles purple tipped. Rhinophores brown with purple tip. Twenty-two irregularly branched gills, fully retractile, spirally arranged around anus.

ASHOD FRANCIS

FAMILY: KENTRODORIDAE: Kentrodorids

Kentrodoris funebris has a white body and black or very dark red spot and circles on the back. Rhinophores deep red in distal portion. Short spiky organs, called caryophyllidia, are arranged on the mantle in blackish-red circles. Blackish spots on foot posteriorly but ventral surface is white.

Asteronotus caespitosa is a large nudibranch, often more than 10cms. Dorsal surface smooth but with a series of tubercles and ridges. A mid-dorsal ridge carries several tubercles and there are large and irregularly arranged tubercles on each side of this ridge. Approximately four ridges run parallel to the outer edge of the mantle. Mantle is blackish-grey while ridges are yellow-brown or grey-brown. Rhinophores brown with white line down posterior surface or stalk. Gills brown; anus white near tip. Foot grey.

FAMILY: DISCODORIDAE

Platydoris scabra has a white mantle with dense brown spots and patches so that in places it appears to be completely brown. Approx. 4-8cms long; flattened, with a minutely granular dorsal surface. In darker areas of the mantle there is a mesh pattern of brown rings with white centres. Mantle edge, tips of oral tentacles and rhinophores, gill sockets and edge of foot are all orange. Ventral surface of foot notched and transversely grooved.

Discodoris fragilis. Collected specimens generally range from 2 to 10cms in length. The foot does not extend posteriorly beyond the mantle. Surface of mantle is covered by low, rounded (occasionally pointed) tubercles. Colour: mixture of brown, dark brown (or purple-brown), buff and cream or white. Mantle edge paler than centre. Ventrally, foot grey with a midline pattern of brown spots.

Sebadoris nubilosa occurs on sponges in shallow water, especially on sea grass beds. Collected specimens generally range from 2 to 14cms. Body dorsoventrally flattened, soft and slimy. Mantle covered by a dense arrangement of papillae. Rhinophores grey, mottled with brown and white (or cream) sharply angled with front of club flattened and with a conspicuous white line running up its rigid mid-line. Body pale grey with complex mottling of various shades of brown, sometimes tinged with red or yellow. Ventral surface of foot notched and transversely grooved; white with scattered brown spots.

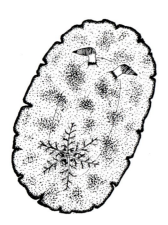

Platydoris scabra

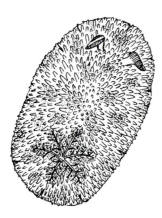

Sebadoris nubilosa

Kentrodoris funebris

JIM DORAN

Order: Aeolidacea

Characterized by bundles or rows of cerata which contain extensions of the digestive gland. The tips of these are modified into special defensive organs, the cnidosacs. Aeolids which feed on nematocyst bearing coelenterates divert their stinging cells to the cnidosacs where they are stored and may be discharged when threatened.

FAMILY: AEOLIDIIDAE

Two pairs of tentacles on head and rows of cerata on back; e.g. *Aeolidiella orientalis* is 1cm long. Yellow-orange, tips of cerata and tentacles whitish.

FAMILY: CUTHONIDAE

Tiny aeolid nudibranchs with a pair of prominent oral tentacles and rhinophores and groups of cerata which carry cnidosacs at their ends.

FAMILY: PTERAEOLIDIIDAE

Pteraeolidia semperi is approx. 3cms long; white with numerous groups of light brown cerata arranged along the full length of its narrow body; each tentacle has two purple rings; tips of rhinophores and tip of tail are purple.

Order: Onchidacea

Oval-shaped with a superficial resemblance to chitons. Single pair of tentacles on head and pair of skin folds on either side of mouth. Extensive mantle frequently with warts on dorsal surface. Sometimes included in subclass Pulmonata since a posterior cavity serves like a lung. Typically live in mangrove flats or among crevices in rocky inter-tidal areas. They can respire in both air and water.

FAMILY: ONCHIDIIDAE

Peronia peronii. Collected specimens generally within the size range 5 to 15cms long. Head short and very broad. No jaws but buccal cavity has horny lining. Radula present.

Pteraeolidia semperi

Pteraeolidia semperi

F. JACK JACKSON

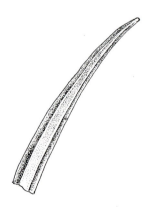

Siphonaria sp.

Subclass: Pulmonata: Lung Shells

Gastropods which lack gills and in which the anterior part of the mantle cavity fulfills the function of a respiratory organ enabling pulmonates to respire in air and under water. Shells display varying degrees of uncoiling (detorsion) from the typically coiled gastropod form to a conical uncoiled shape. In some species the shell is reduced or else completely absent.

Order: Basommatophora

Shelled aquatic pulmonates with single pair of tentacles which carry an eye at their base. Spirally coiled or limpet-like shells, usually found intertidally.

FAMILY: SIPHONARIIDAE

Members of this family have a limpet-like shell, somewhat flattened but with a pulmonar groove inside.

Siphonaria kurracheensis is quite common on rocks and boulders in the inter-tidal zone where it grazes on micro-algal lawns which cover hard substrata.

FAMILY: ELLOBIIDAE

Superficial resemblance to cone-shells but the aperture is toothed. Represented species include: *Melampus lividus* and *Melampus flavus*.

Class: Scaphopoda: Elephant's Tusk Shells

Marine group of molluscs whose shells resemble the shape of an elephant's tusk. The gently curved, narrow, tubular shell is open at each end. A functional diagram is given in figure 33.

FAMILY: DENTALIIDAE

These live in silt or sand with only the narrow posterior tip of their shells protruding above the sediment.

Dentalium elephantinum has a longitudinally ribbed shell which is dark green towards the anterior and lightening to white posteriorly. Generally found in mud or silt.

Dentalium longirostrum has a translucent shiny amber shell which is slim and smooth. Occurs in mud or fine silt.

Dentalium bisexangulatum has a thin white shell, round in cross-section, with twelve longitudinal ribs. It occurs in relatively coarse lagoon sand.

Dentalium reevei has nine primary ribs with three intermediate riblets.

Dentalium anatorum is probably a Red Sea endemic species which has ten to thirteen ribs and three intermediate riblets.

Figure 33:
Functional diagram of Scaphopod mollusc

Dentalium reevei

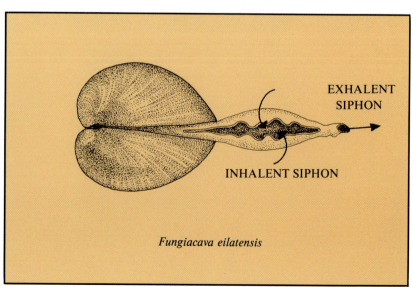

Fungiacava eilatensis

Class: Bivalvia: Bivalves

Bivalve molluscs include the familiar cockles, mussels, clams, oysters and scallops. The two shell valves are hinged by an elastic ligament which tends to open the shell valves. The strong closing force is provided by a pair of adductor muscles which are firmly attached to the inside of the shells and which leave prominent scars on the inner shell surface after the animal has been removed. Generalized bivalve features are illustrated in figure 34.

FAMILY: ARCIDAE: Ark Shells

Solid, radially ribbed, shelled bivalves, often covered by a hairy periostracum and usually with a projection of the shell posteriorly. Both valves similar in shape and normally trapeziform. Most species are of moderate size (e.g. 5cms) and have a characteristic straight hinge bearing many comb teeth.

Arca ventricosa: Ventricose Ark
This has a zebra-like pattern of dark brown stripes in the wide ligamental area. The shell surface has rough radial ribs and fine concentric striations.

Anadara antiquata: Antique Ark
Uniform ribs but without a distinct posterior ridge. Not uncommon in mud and fine sediments or among sea-grass flats.

Arca plicata: Plicate Ark
Lives attached by its byssus to rocks in the littoral zone or under coral heads and in crevices sublittorally. Its shell is white to yellowish with conspicuous oblique ribs posteriorly, and a cross pattern formed by concentric striations and ribs in the mid-shell and anterior portion.

FAMILY: GLYCYMERIDAE: Bittersweet Clams or Dog Cockles

Shells thick and of medium size (e.g. 2-10cms long) with each valve almost circular. Smooth or ribbed and often brightly coloured. Beak in mid-line; hinge-line straight and teeth in curved row. Two adductor muscle scars approximately equal.

Glycymeris pectunculus: Comb Dog Cockle
This has a shell pattern which is formed by concentric zig-zag bands of brown or black.

FAMILY: LIMOPSIDAE

Major characteristics are similar to Glycymeridae. Represented species include *Limopsis multistriata* which occurs in fine sand, especially among sea-grasses in shallow water.

Order: Mytiloida

FAMILY: MYTILIDAE: Mussels

Generally elongate, brown to purplish black shells which possess a hairy periostracum. Byssal threads are used to attach the mussel to a hard substratum. They are filter feeders and include some coral boring species.

Fungiacava eilatensis. This small mytilid is an atypical form which burrows into the skeleton of living fungid corals by means of a chemical secreted by the epithelium of the pallial envelope surrounding the extremely delicate shell. The same epithelium secretes aragonite and thus fills the cavity which it occupies. The mushroom coral does not exhibit any reaction to the presence of the boring mussel.

Figure 35 shows diagrams of various lithophagous burrows. *Lithophaga cumingiana* bores into *Stylophora* coral. *Lithophaga hanleyana* burrows in several corals including *Cyphastrea, Montipora,* and *Goniastrea* species, while *Lithophaga lima* also occurs in *Montipora. Lithophaga teres* bores into a number of shallow-water corals. It has a dark brown shell. *Modiolus auriculatus* is a free-living mussel which has a thin shell, blue on the inside and brownish externally. It occurs intertidally, attached to coral boulders.

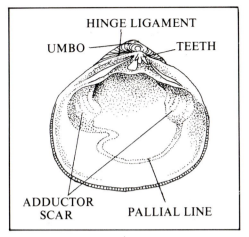

Figure 34: Generalized bivalve

Anadara antiquata

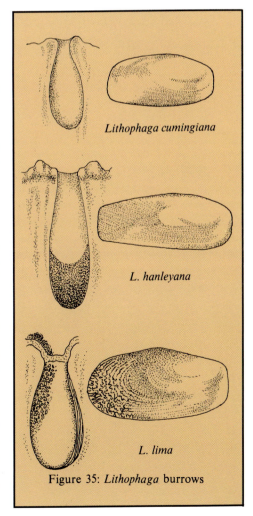

Figure 35: *Lithophaga* burrows

Date mussels: *Lithophaga* sp. in *Diploastrea* coral.

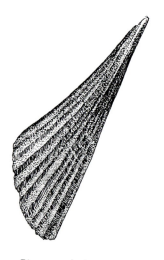

Pinna muricata

FAMILY: PINNIDAE: Razor Pen Shells

Elongate, triangular, brittle shells which usually have an inner nacreous surface. Pen shells normally live with most of their shell embedded in sand or gravel where it is anchored by long byssal threads. The protruding posterior edge of shell is very sharp. Represented species in the Red Sea include:

Streptopinna saccata: The Baggy Pen Shell.
This has a fragile thin shell (approx. 20cms long) of irregular shape; fused along the anterior edges and only open at the posterior end. It is usually found in association with living corals.

Pinna muricata: Prickly Pen Shell.
This species is more typical of the pen shell family, having a fan-shaped shell (approx. 15cms) which has about twelve low ridges which, in younger forms, carry short spines in their posterior section. It is a relatively common species on sea-grass beds in shallow water.

Pinna bicolor has a thin, paddle-shaped shell with the posterior edge straight. It lives embedded in sand and attached by its byssal threads.

Atrina vexillum has a triangular shell (up to 30cms long) which is usually found embedded among corals. The shell interior is glossy black.

Order: Pteroidea

Superfamily: Pteriacea

FAMILY: PTERIIDAE: Pearl Shells

Members of this family have shells with a well developed, lustrous nacreous lining internally, and they are therefore in demand as the raw material for mother-of-pearl. The valves are joined by a long hinge line which has a small anterior ear-like projection and an expanded posterior projection. Represented species include:

Pinctada margaritifera: Black-lip Pearl Shell
This species is the main pearl shell in the Red Sea and is of particular interest as a result of its artificial cultivation at Dunganab Bay, along the northern coast of Sudan, where the species undergoes a well defined spawning season which has resulted in the development of techniques for collection of large quantities of their spat (between June and September). Natural pearls are rare in this species but artificial pearl culture is possible by implantation of nuclei (usually made from plastic) into the shell. In the wild this species grows to around 25cms and is found attached to coral rocks or living free on the sea-bed.

Below: *Pinctada margaritifera;* right: *Pteria aegyptiaca* on soft coral.

Pinctada radiata: The Rayed Pearl Oyster
This has a smaller (e.g. 6cms) red-brown shell with several radiating bands of a lighter colour.

Electroma alacorvi: Raven's Wing
Has a thin, brittle shell (approx. 6 cms) which is normally protected as a result of its habitat; i.e. living embedded within corals with only the shell edge visible.

Pteria penguin: Penguin's Wing
Black shell with a bluish pearly interior and a long posterior projection (or wing) extending from the hinge-line. Up to about 25cms. Found at moderate depths, often attached to horny corals.

FAMILY: ISOGNOMONIDAE: Toothed Pearl Shells
These are superficially similar to pearl oysters but their shell is irregular in shape, flattened, and elongate. The straight hinge line bears flattened simple teeth. Externally, their valves tend to be flaky while the internal surface has a nacreous lustre.

Represented species include *Isognomon legumen:* the Pod Tree Oyster, which has a thin cream or yellowish tongue-shaped shell (approx. 9cms).

FAMILY: MALLEIDAE: Hammer Oysters and Sponge Fingers
These are characterized by a hinge which is devoid of teeth. Shell valves are elongate vertically (i.e. from hinge to foot). The family includes finger-shaped forms which are often associated with sponges *(Vulsella* spp) and hammer-shaped bivalves *(Malleus* spp). Represented species include:

Vulsella vulsella: the Sponge Finger Oyster, which has a dark brown to yellowish shell which may have darker stripes (approx. 7cms).

Pteria penguin

Vulsella vulsella

Pteria sp. on gorgonian

ASHOD FRANCIS

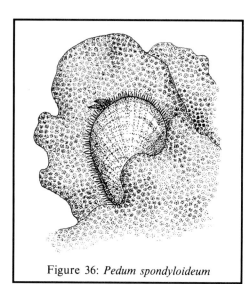

Figure 36: *Pedum spondyloideum*

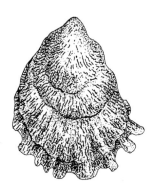

Plicatula plicata

Spondylus aurantius

Superfamily: Pectinacea
FAMILY: PECTINIDAE: Scallops

These well-known forms have fan-like shells with rayed ribs. The right valve is usually deeply concave while the left one is more or less flat. There are projections at each end of the hinge line. Scallops are able to swim by a process of jet-propulsion; water taken in as the valves open is rapidly expelled on each side of the hinge line as the valves are rapidly clamped closed. This activity is triggered by the approach of predators which scallops are able to sense. Represented species include:

Pedum spondyloideum (figure 36) which is somewhat atypical of the family as a result of the fact that it is invariably associated with living corals (frequently *Porites*). It is often overlooked by shell collectors since it lives very firmly attached within crevices on corals. The right valve is concave with a deep byssal notch while the left, upper one is almost flat. The outer surface of the right valve has fine concentric lines while the left valve has prominent radiating ridges with fine spines distally. Both surfaces are white with scattered purple blotches. *Pedum* does not bore into coral but inhibits coral-growth in its immediate vicinity which results in the development of crevices as the rest of the coral grows.

Gloripallium sanguinolenta: Blood-stained Scallop
Has a white strong, radially ribbed, shell (approx. 6cms) with red, orange or violet spots. Its surface is marked with fine growth increments. It occurs among coral debris in the fore-reef area.

Chlamys squamosa: The Squamose Scallop
Has a darkish brown to purple shell (approx. 6cms) which bears fifteen to twenty low radiating ridges on which there may be lighter coloured scales. It is found in association with living corals.

FAMILY: PLICATULIDAE: Plicate Oysters or Kitten's Paws

Related to thorny oysters and to scallops, the Plicate Oysters have relatively small shells (e.g. 4.5cms) in which the upper valve bears tile-like projections on radiating ribs. Hinge is short but teeth are prominent. They are generally white to tan coloured with a pattern of speckled brown spots. Their right valve is attached to the substrate. Represented species include:

Plicatula plicata: The Plicate Kitten's Paw
This species has a well formed, roughly triangular ribbed shell (approx. 5cms) which is generally cream with a pattern of radiating brown lines or bands.

FAMILY: SPONDYLIDAE: Thorny Oysters

These large scallop-like shells have one cup-shaped valve firmly attached to the substrate, while the other valve is flat and acts like a lid. This upper valve usually bears long spines which emanate from radial ribs. The oysters generally occupy prominent situations on the fore-reef, in full view of all comers, and their spiny armament is no doubt an effective deterrent to potential predators. Represented species include:

Spondylus marisrubri: The Red Thorny Oyster
Shell (10cms) is reddish brown with spines which may be similarly coloured or else white.

Spondylus aurantius: Golden Thorny Oyster
Shell large (up to 20cms) with right valve deeply concave. Golden yellow to orange with long white spines and a single row of finer spines between. Often found attached to rock under shaded overhangs. Valves may often be covered by a red sponge.

Superfamily: Anomiacea
FAMILY: ANOMIIDAE: Jingle Shells

These flattened oyster-like molluscs are characterized by a hole in one valve through which the byssus passes. The valves are thin, irregularly shaped and translucent. Represented species include:

Anomia nobilis which is found attached to the underside of coral boulders in the littoral zone or in very shallow areas.

Superfamily: Limacea
FAMILY: LIMIDAE: File Shells
In life, file-shells are quite spectacular since the animal is bright red or orange and has long sticky tentacles which protrude from its gaping valves. They occur under dead coral boulders on sand flats or among coral and shell rubble. They have the ability to swim. Represented species include:

Lima lima: Spiny File Shell
It has a white shell with eighteen to twenty six closely spaced scaly ribs. It is found in coarse sand and coral debris in fore-reef.

Superfamily: Ostreacea
FAMILY: GRYPHAEIDAE: Honeycomb Oysters
The shells of these oyster-like molluscs have a cellular layer sandwiched within the hard layers. Represented species include:

Hyotissa hyotis, which has a heavily constructed shell bearing fifteen to twenty sharp ribs with long tubular spines.

FAMILY: OSTREIDAE: True Oysters
Oysters live with their lower (right) valve firmly cemented to the substrate. They have a simple hinge-line which lacks teeth. The inside of the shell is chalky or lustrous and has a single adductor muscle scar. Represented species include:

Lopha cristagalli, which has an unmistakable shell (10cms) with a zig-zag edge formed by several prominent angular ridges. It is often found attached to black coral trees in moderate depths. Its valves are frequently encrusted by the red sponge *Microciona* sp.

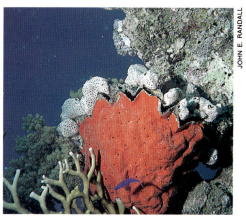

Lopha cristagalli

Thorny oyster: *Spondylus* sp.

Subclass: Heterodonta

This large subclass includes many familiar bivalves such as cockles, venus shells, giant clams and boring shipworms.

Order: Veneroida

Bivalves with similar shell valves. Beaks normally towards anterior end. Hinge with two types of teeth. Two equal sized adductor muscles. Siphons are often present. Includes species which live on the surface of sediment as well as burrowing or attached forms.

Superfamily: Lucinacea

FAMILY: LUCINIDAE: Saucer Shells

More or less circular and flattish shells with similar chalky white valves which are patterned with concentric and radiating grooves. Two central cardinal teeth at hinge of each valve together with a posterior lateral tooth on hinge of right valve which fits into a corresponding depression on the left valve hinge. Foot elongate and used for burrowing in sand or coral gravel. Represented species include:

Codakia tigerina: Tiger Lucine
Radially ribbed shell with numerous concentric striations and more prominent growth lines. White on outside and yellowish with reddish margin on inside. It occurs in coarse sand and debris in lagoons or in the fore-reef area.

Lucina dentifera: Toothed Lucine
Shell with prominent growth rings and a radial groove posteriorly. It is found attached to rocks and boulders in the littoral zone.

FAMILY: UNGULINIDAE

These are smooth, rounded lucinid shells. Represented species include *Diplodonta rotundata*.

Superfamily: Chamacea

FAMILY: CHAMIDAE: Chamas or Jewel Box

With unequal shell valves, one of which is cup-shaped and cemented to the substrate while the upper valve acts like a lid. Scaly or spiny projections are often present. They live attached to coral boulders or in crevices on the reef-platform. Represented species include:

Chama pacifica which has a red-orange shell (approximately 7.5cms) with distinct short blunt white spines.

Superfamily: Carditidae

FAMILY: CARDITIDAE: Cardita Clams

Thick shells with prominent radial ribs. White to dark brown and variegated. Hinge line carries oblique interlocking teeth, usually with two teeth on one valve and a single tooth on the opposite valve. Two adductor scars are almost equal and the pallial line is not indented. Represented species include:

Beguina gubernaculum: The Rudder Cardita
Somewhat flattened shell with radial ribs which bear scales on the posterior ones. Brown to red colour (approx. 6cms).

Superfamily: Cardiacea

FAMILY: CARDIIDAE: Heart Cockles

Shells are swollen and frequently heart-shaped. There are four teeth in the hinge-line. Valves are radially ribbed and in some cases ribs carry spines or scales. The shell margins are crenulated so that the valves interlock when closed. Heart cockles are efficient burrowers in sand or mud and they have a large foot which is used for this purpose. The two siphons are quite short and thus dictate the level underneath the sediment which the cockles occupy. Represented species include:

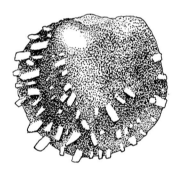
Codakia tigerina

Chama pacifica

Fragum (Lunulicardia) auricula

Fragum (Lunulicardia) auricula: the Ear Cockle, which has a sharp posterior ridge and rounded ribs. It is white with a scattering of red spots (approx. 5cms).

Nemocardium lyratum has a rounded shell (4-6cms) with radial ribs which are superimposed by oblique secondary ribs. It has a red periostracum with a white beak and white spots which result from erosion of the red layer. Occurs in coarse sand of lagoons and in the fore-reef.

Superfamily: Tridacnacea
FAMILY: TRIDACNIDAE: Clams
   Generally large, solid shells in which each valve has an undulating external form resulting from a number of radiating ribs. In many species these carry scales. They frequently occur among living corals. Represented species include:

Tridacna crocea: The Crocus Giant Clam
This bores into corals and dead coral boulders. It has a thick shell with about six rounded ribs which are adorned by a concentric series of large open fluted scales. It is a relatively small clam species with a shell length which rarely exceeds 15cms. Although a filter feeder, it derives much of its nutritional requirements from the photosynthetic activity of the single celled algae (zooxanthellae) which are embedded in its mantle.

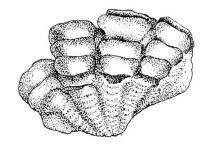

Tridacna crocea

Tridacna squamosa: Fluted Giant Clam
This is larger than the previous species. Its ribs are more widely spaced and they bear open, tubular fluted scales which have a more separate appearance (between adjacent ribs) than those of *T. crocea.* Also depends on zooxanthellae in its mantle for a major part of its nutritional requirements.

Tridacna maxima has an elongate, triagonal shell with a dense arrangement of foliaceous hollow scales which are arranged in folds so that the margin has a wavy appearance. Like the above two species it carries zooxanthellae in its mantle and these play a significant role in its nutrition. Despite its scientific name it should not be confused with the Indo-Pacific Giant Clam, *Tridacna gigas,* whose shell length can exceed one metre. This species is rarely longer than 40cms.

Tridacna maxima

J. DAVID GEORGE

Superfamily: Mactracea

FAMILY: MESODESMATIDAE: Surf Clams or little Wedge Shells

Small wedge-shaped shells (2-3cms) with solid valves each carrying a single tooth. Smooth or with concentric ridges. Siphons separated from each other. Live intertidally, in sand. Represented species include:

Atactodea glabrata: Smooth Beach Clam
Somewhat triagonal, thick shell (3cms) in which the pallial sinus is a U-shaped indentation. It is not uncommon in mud and fine sediments.

FAMILY: MACTRIDAE: Trough Shells

Shell usually swollen and triagonal. Each valve has two cardinal and two lateral teeth on its hinge-line. Externally the valves are relatively smooth and concentrically grooved. Represented species include *Mactra olorina,* which occurs in mud and fine sediments.

Superfamily: Tellinacea

FAMILY: TELLINIDAE: Tellins

Frequently brightly coloured bivalves in which each valve is compressed and bears fine concentric growth striations sometimes accompanied by radial grooves. Hinge line with poorly developed teeth but a conspicuous ligament externally. Tellins are able to burrow quite deeply into sand and they have a broad foot and long siphons, which permit them to maintain water circulation while they are several centimetres under the sand. They feed by sweeping the sediment surface with their inhalent siphon. Representatives include:

Tellina virgata which has a strong elongate shell (6-8cms) with numerous concentric striations. White to yellow with red or purple radial rays which broaden towards the ventral margin. Lives in mud and fine sediments sublittorally.

Tellinella staurella: The Cross Tellin.
This species lives in coarse lagoon sand; approx 6cms.

Scutarcopagia scobinata: The Rasp Tellin
Occurs in quite a wide range of habitats from shallow sandy areas to deeper regions of the fore-reef. Its shell has a rasp-like surface.

FAMILY: DONACIDAE: Wedge Clams

Generally small shells which may be wedge-shaped and have a pattern of radial striations. The posterior end is oblique. They usually live just beneath the surface of the sand in shallow areas where wave turbulence occurs. If dislodged by waves they are adept at re-burying themselves under the sediment with the aid of an exceptionally agile foot. Represented species include: *Donax erythraeensis.*

FAMILY: SEMELIDAE

Similar in general form and habitat to tellins. They live in fine sediment or mud. Represented species include *Leptomya rostrata.*

FAMILY: PSAMMOBIDAE
Sanguin Clams, Gari Clams, Sunset Clams

Some species have coloured rays which originated at the umbo and diverge to produce a radiating pattern reminiscent of a sunset. Elongate, compressed shells which gape posteriorly. Teeth are weakly developed (lateral teeth absent) but a conspicuous external ligament is present. Shell valves are usually quite thin and have numerous concentric striations. They have long siphons and burrow deeply into coral sand or sandy mud surrounding coral reefs. Represented species include:

Asaphis violascens: The Violet Asphis
Shell approximately 6cms with coarse radial sculpturing and concentric lines. The fore and hind regions are lightly scaled: Pink, violet or yellow. Occurs in soft sediment in shallow water to moderate depths.

Atactodea glabrata

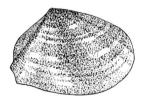

Tellinella staurella

Asaphis violascens

FAMILY: SOLECURTIDAE: Solecurtid Clams

Elongate bivalves which gape at each end. Their shell has a medial depression and obliquely diverging striations sculpture the external surface. Anterior and posterior adductor scars are of roughly similar size and pallial line is deeply indented. Represented species include: *Solenocurtus australis:* the Southern Solecurtus (approx. 6cms).

Superfamily: Arctiacea

FAMILY: TRAPEZIIDAE: Trapezium Clams

Oblong clams in which the hinge line is curved and bears strong teeth including a large blade-like posterior lateral tooth. Adductor muscle scars are dissimilar in size, the anterior one being smaller. Pallial line indented by shallow sinus. Live attached by byssus to hard substrates. Represented species include:

Trapezium oblongum: The Oblong Trapezium
Inflated shell with a prominent posterior ridge which becomes more rounded with age.

Superfamily: Veneracea

FAMILY: VENERIDAE: Venus Clams

Frequently strikingly coloured bivalves which are strongly sculptured. The curved, eccentric hinge carries strong teeth including three to four cardinal teeth and one lateral tooth on each valve. Adductor scars are of approximately equal size. External ligament and prominent lunule on hinge line. Strong foot for burrowing and siphons are fused along most of their length. Generally found in coral sand, just beneath the surface. Represented species include:

Irus macrophyllus: The Long Leaf Irus
Small shell (approx. 2cms) truncated posteriorly. Growth lamellae and radial ridges. It is found under coral heads, in crevices and among coral rubble where it tends to bury into sand pockets.

Tapes sulcarius: The Furrowed Venus
Cream with irregular brown lines and three incomplete dark brown rays (approx. 6cms). It lives in mud or fine sediments in relatively shallow water.

Venus lamellaris has a somewhat triangular shell (4-5cms) with prominent concentric costae and radial ribs. Creamy white with brown spots. Lives in sand from the upper reef to at least 50m.

Circe scripta has a rounded shell (4-5cms) with a broad beak. The surface is sculptured with numerous concentric grooves. Colour pattern consists of a white background with irregular brown markings which have the appearance of writing, hence its Latin name.

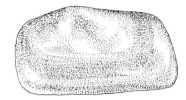

Solenocurtus australis

Trapezium oblongum

Irus macrophyllus

Lamellibranch: *Gastrochaena* sp.

FAMILY: PETRICOLIDAE: False Angel Wings

Elongate venerid shells which bore into soft coral rock. The hinge and umbo is towards the anterior end. Their common name results from their appearance when both valves are opened out. Representatives include: *Petricola hemprichi.*

Superfamily: Leptonacea

FAMILY: GALAEOMMATIDAE: Galeommas

These tiny bivalves have thin, compressed shells with minute teeth. They occur among live corals. Represented species include: *Galeomma mauritiana:* The Mauritian Galeomma (approx. 1cm).

FAMILY: MONTACUTIDAE

Small Leptonaceds, generally similar to the previous family. Represented species include *Montaguia lamellifera.*

Superfamily: Crassatellacea

FAMILY: CRASSATELLIDAE

Carditoid family whose species generally possess thick rounded shells with concentric ribs. Represented species include *Eucrassatella jousseaumei.*

Superfamily: Solenacea

FAMILY: SOLENIDAE: Finger Oysters or Jack-knife Clams

Long, narrow shells which are open at the posterior end. Thin shell valves are usually smooth and may be straight or curved. Hinge line carries a pair of cardinal teeth and an external ligament is present. They have a well developed foot which is an extremely efficient burrowing organ. They usually live in muddy sand or silt, particularly in mangrove areas and among sea-grasses. Represented species in the Red Sea include *Solen truncatus.*

FAMILY: CULTELLIDAE: Razor Shells

Similar to the previous family. Shell valves are long and rectangular. Represented species include:*Cultellus cultellus,* which has a thin curved, elongate shell (approx. 5cms) which is translucent, olive coloured with reddish brown spots.

Order: Myoida

Bivalves which burrow into mud, coral, rock or wood. Thin, generally dissimilar shell valves. Hinge has few or no teeth and ligament often reduced causing shell to gape open. Ventral edges of mantle fused except where foot protrudes. Long, well developed siphons are present.

Superfamily: Myacea

FAMILY: MYIDAE: Soft Shell Clams

An ovoid shell, open posteriorly, with long siphons which cannot be withdrawn and lie in a wrinkled tube formed by an extension of the periostracum. Represented species include *Cryptomya decurtata.*

Superfamily: Gastrochaenacea

FAMILY: GASTROCHAENIDAE: Flask Shells

These have no teeth on the hinge-line. They live burrowed into soft rocks or corals. A bottle-like extension from the burrow has a narrow opening in the shape of a figure of eight which provides access to the outside of the siphons.

Gastrochaena cuneiformis has a characteristically shaped shell (up to 3cms) and lives embedded in coral or soft rock with only its siphons visible.

Superfamily: Pholadacea

FAMILY: PHOLADIDAE: Angel's Wing Shells

These forms bore into rock, coral and wood. Their long, swollen

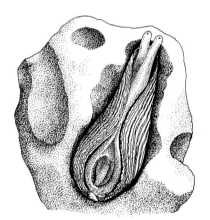

Pholas dactylus

Gastrochaena sp.

shells gape at each end. Valves have a sculptured pattern formed by concentric ridges and radiating lines. The hinge has a peg-like internal projection. The mantle edges are fused except where the foot protrudes. Two elongate siphons are also united. Represented species include:

Pholas dactylus, which has an elliptical shell (8-15cms) pointed anteriorly and rounded at the posterior. Sculptured with concentric and radial ribs. White to grey. Bores into wood or soft calcareous rock in the shallows.

Subclass: Anomalodesmata

Shell valves often elongate and always dissimilar to each other. Hinge teeth reduced or absent. Adductor muscles are usually approximately equal in size (occasionally the anterior muscle is smaller). Mantle edges fused and long siphons joined together and sometimes enclosed in the calcareous tube.

Order: Pholadomyoida

Superfamily: Pandoracea
FAMILY: LATERNULIDAE: Lantern Shells

These shells lack teeth. Umbone has a natural split. They burrow in mud or fine sand. Represented species include *Laternula anatina:* the Duck Lantern Shell (approx. 7cms).

Laternula anatina

FAMILY: LYONSIIDAE

Related to the above family and with similar features. Represented species include *Lyonsia intracta.*

Superfamily: Poromyacea
FAMILY: CUSPIDARIIDAE

Represented species include *Cuspidaria dissociata,* which was originally collected by dredging near Jeddah.

Superfamily: Clavagellacea
FAMILY: CLAVAGELLIDAE: Watering Pots

Long, fragile tubular shells (approx. 15cms). Live buried in sand with open end just extended above surface of sediment. Represented species include:

Brechites attrahens: the Furbellowed Watering Pot, which has a long, tubular, white shell. The closed end is perforated like a sieve and surrounded by finger-like projections. The open end is ruffled and in some cases there are several extensions.

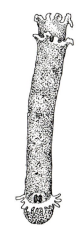

Brechites attrahens

Tridacna sp.

F. JACK JACKSON

HORST MOOSLEITNER

Class: Cephalopoda: Octopus, Squid and Cuttlefish

Subclass: Coleoidea

Cephalopods with shells internal or absent. Head with eight to ten suckered tentacles and a pair of well developed eyes. Funnel always a closed tube and single pair of gills in mantle cavity. Most species possess an ink sac and skin with pigment cells.

Order: Sepioidea: Cuttlefish and Bottle-tailed Squids

Cephalopods with internal calcareous shell which is coiled and chambered or straight; or reduced and horny; or absent altogether. Body usually short and broad with lateral fins. Head carries ten tentacles, two of which are longer and more mobile than the others and have suckers only at their distal, spoon shaped ends.

FAMILY: SPIRULIDAE: Spirula Squid

Spirula spirula is a common, widely distributed deepwater squid with a cylindrical body which is rarely more than 7cms long, and which has a spirally coiled internal shell visible at the posterior end. They shoal in open oceanic water, at depths from 100m to 2000m and feed on a variety of planktonic organisms.

FAMILY: SEPIIDAE: Cuttlefish

These have a spongy calcareous internal shell. The body has a long narrow fin on each side. There are ten arms, two of which are longer than the others and are modified for capture of prey. Cuttlefish are usually seen at night, when they are attracted to lights.

Sepia dollfusi (Cuttlefish) is smaller than *Sepia pharaonis*, and its internal shell seldom exceeds 10cms. It has a prominent projection of the mantle in the mid-dorsal anterior region. Arms with four rows of suckers; club tentacles longer (approx. 35% of the length of the mantle). The shell is oval, without a point at the posterior. On the dorsal side it has a weakly developed median ridge. With the exception of a narrow edge it is completely calcified.

Sepia pharaonis (Cuttlefish) is similar to the above species but larger, with cuttlefish bone up to 20cms in length and with a mid dorsal ridge and relatively wide lateral extensions (chitinous) and a pointed posterior end. Other recorded *Sepia* species include: *Sepia savignyi; Sepia gibba; Sepia elongata; Sepia trygonina; Sepia australis* and *Sepia prashadi*.

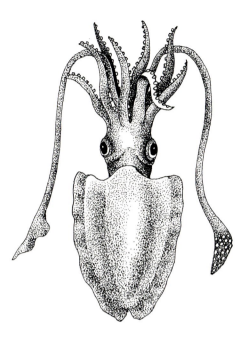

Cuttlefish

Opposite: *Octopus macropus*
Sepioteuthis lessoniana

HORST MOOSLEITNER

Sepioteuthis lessoniana

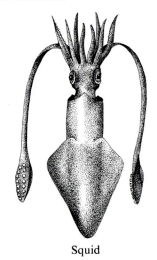

Squid

Octopus cyanea

Order: Teuthoidea: Squids

Cephalopods with an internal horny, long pen-shaped shell. The body is elongate and often torpedo-shaped, with lateral fins. Head with ten tentacles, two of which are usually longer than others.

FAMILY: LOLIGINIDAE: Squids

Sepioteuthis lessoniana has an elongate body, fins extended along the entire length of the mantle and are quite wide (approx. 18% of the mantle length) with the greatest width posterior to the middle. Smaller specimens have maximum fin width more in line with the middle of the mantle than do larger ones.

FAMILY: ENOPLOTEUTHIDAE

Small squid with relatively large fins and with light organs.

Abralia steindachneri is a relatively large member of the family with a mantle length of about 5cms. Fins rhombiform. Short arms approximately half the mantle length. Several large luminous organs in the ocular bulbs; the long arms each carry three to four rows of luminous organs with a protective membrane.

Enoploteuthis dubia is also recorded from the Gulf of Aqaba.

FAMILY: OMMASTREPHIDAE

Mantle long and narrow, posterior pointed; fins triangular; the horny pen is very thin. Represented species include:

Ilex coindetii. Fins short, hardly more than a third of the length of the mantle which is about 25cms long.

Symplectoteuthis oualensis. Fins are approximately half the mantle length which is about 28cms long.

Ommastrephus arabicus is another representative of this family which occurs in the Red Sea.

FAMILY: CHIROTEUTHIDAE

Mantle slim with the pointed posterior section extending beyond the fins; often with luminous organs; tentacles very long with rows of luminous organs along their external edge. Represented by the genus *Chiroteuthis*.

Order: Octopoda: Octopus

Body generally round and sac-like without fins; internal shell rudimentary or absent; mantle with a curved indentation on its dorsal edge behind the head; eight arms, joined by a membrane, with suckers in one or two rows.

FAMILY: OCTOPODIDAE

Octopus cyaneus is usually a deep red-purple colour or blackish, with two large eye spots between the eyes and the edge of the interbrachial membrane. The mantle length is about 12cms and overall length can reach about 300cms. There is a double row of suckers on the arms.

Other members of the family recorded from the Red Sea include: *Octopus vulgaris; Octopus rugosus; Octopus macropus; Octopus horridus; Octopus robsoni; Octopus aegina* and *Eledone moschata*.

Octopus cyanea

FAMILY: TREMOCTOPODIDAE

Oceanic octopus which have dorsal arms much longer than the others and with all the arms united by a large strong membrane so that a net is created and used for capture of its prey.

Tremoctopus violaceus. Adult females measure about 12cms along the mantle whereas males are much smaller with a mantle length of only 1.5cms or so. The right ventro-lateral arm of male is transformed into a complex hectocotyl organ.

FAMILY: ARGONAUTIDAE

Interbrachial membrane is poorly developed. The dorsal arms of the female secrete a secondary shell for incubation of her eggs.

Young octopus

Argonauta argo: Paper Nautilus
This oceanic species is rarely encountered in the Red Sea but it does occur there. The female secretes a thin protective shell by means of modified glandular arms. She uses this for brooding her fertilised eggs and for protection of the adult. Smaller males do not secrete such a shell. Mantle length of females is about 10cms. Males are about one eighth this size and are characterized (apart from their small size) by a specially adapted hectocotylous arm which is used for mating. The female shell is particularly attractive and is in the form of a white, fragile, coiled chamber. The shell has a keel formed from a double series of tubercles.

Argonauta hians is also present and is distinguished from the previous species by having a larger shell keel formed by stronger but less numerous tubercles.

Octopus cyanea

Table XVII.

RED SEA MOLLUSCS

ACANTHOCHITONIDAE

Acanthochiton curvisetosus	Craspedochiton laqueatus
Acanthochiton discrepans	Cryptoplax sykesi
Acanthochiton fascicularis	Notoplax aqabaensis
Acanthochiton mastalleri	Notoplax elegans
Acanthochiton penicillatus	

ISCHNOCHITONIDAE

Ischnochiton yerburi	Callistochiton heterodon

CHITONIDAE

Chiton canariensis	Tonicia dilecta
Chiton olivaceus affinis	Tonicia perligera
Chiton platei	Tonicia scabiosus
Choneplax parvus	Tonicia suezensis
Tonicia costata	Acanthopleura haddoni

HALIOTIDAE

Haliotis dringi	Haliotis unilateralis
Haliotis pustulata	Haliotis varia

FISSURELLIDAE

Diodora rueppelli	Hemitoma imbricata
Emarginula arabica	Hemitoma tricarinata
Emarginula arconatii	Clypidina notata
Emarginula cuvieri	Scutus unguis
Emarginula retecosa	Nesta candida
Emarginula rugosa	Macroschisma compressa
Emarginula spinosa	Macroschisma sinensis
Emarginula thomasi	

SCISSURELLIDAE

Scissurella d'orbignyi	Scissurella rota
Scissurella jacksoni	Sinezona armillata
Scissurella reticulata	Sinezona tricarinata

PATELLIDAE

Cellana rota	Cellana eucosmia

ACMAEIDAE

Acmaea saccharina	

TROCHIDAE

Margarites biangulosus	Clanculus pharaonis
Euchelus atratus	Isanda hemprichi
Euchelus bicinctus	Calliostoma fragum
Euchelus pullatus	Agagus agagus
Euchelus scrobiculatus	Trochus bellardii
Monodonta canalifera	Trochus dentatus = Tectus
Monodonta dama	Trochus erythraeus
Monodonta obscura	Trochus hornungi
Turcica stellata	Trochus maculatus
Gibbula declivis	Trochus nodulifer
Gibbula phasianella	Trochus virgatus
Gibbula sepulchralis	Isanda holdsworthiana
Minolia caifassii	Monilea pantanelli
Minolia gradata	Umbonium striolata
Minolia nedyma	Umbonium vestiarium
Clanculus corallinus	Umbonella sismondae

STOMATELLIDAE

Stomatella auriculata	Stomatia phymotis
Stomatella doriae	Gena callosa
Stomatella nigra	Gena varia
Stomatia duplicata	

CYCLOSTREMATIDAE = SKENEIDAE

Cyclostrema biporcata	Cyclostrema micans
Cyclostrema cingulata	Cyclostrema pentegoniostoma

TURBINIDAE

Turbo argyrostomus	Turbo eroopolitanus
Turbo arsinoensis	Turbo petholatus
Turbo coronatus	Turbo pustulatus
Turbo elegans	Turbo radiatus

NERITIDAE

Nerita albicilla	Nerita undata
Nerita neritopsoides	Nerita yoldii
Nerita plexa	Smaragdia feuilleti
Nerita polita	

NERITOPSIDAE

Neritopsis radula	

LITTORINIDAE

Littorina scabra	Nodilittorina subnodosa
Nodilittorina millegrana	Peasiella isseli

RISSOIDAE

Rissoina ambigua	Rissoina seguenziana
Rissoina bertholleti	Rissoina sismondiana
Rissoina bouvieri	Rissoina smithi
Rissoina clathrata	Rissoina spirata
Rissoina ephamilla	Seminella fusiformis
Rissoina erythraea	Alvania orbignyi
Rissoina miltozona	Zebina tridentata
Rissoina nivea	Diala semistriata
Rissoina oryza	Fenella cerithina
Rissoina plicata	Fenella virgata
Rissoina rissoi	Amphitalamus elspethae

ARCHITECTONICIDAE

Solarium perspectivum =	Heliacus areola
Architectonica	Heliacus variegatus
Philippia radiata	

TURRITELLIDAE

Turritella auricincta	Turritella sanguinea
Turritella cochlea	Turritella terebra
Turritella columnaris	Turritella torulosa
Turritella maculata	

CAECIDAE

Caecum annulatum	Caecum arabicum

VERMETIDAE

Vermetus eruca	Dendropoma maxima
Serpulorbis inopertus	

MODULIDAE

Modulus tectum	

CERITHIIDAE

Cerithium caeruleum	Cerithium rueppelli
Cerithium columna	Cerithium scabridum
Cerithium crassilabrum	Cerithium sinensis
Cerithium echinatum	Cerithium subulatum
Cerithium nesioticum	Plesiotrochus souverbianus
Cerithium nodulosum	Clypeomorus bifasciatus
erythraeonense	Clypeomorus concisus
Cerithium moniliferum	Clypeomorus morus
Cerithium pauxillum	Clypeomorus tuberculatus
Cerithium pictum	Rhinoclavis fasciata = Cerithium
Cerithium proteum	Rhinoclavis aspera
Cerithium rarimaculatum	Rhinoclavis kocki
Cerithium rostratum	

POTAMIDIDAE

Pirenella cailliaudi	Royella sinon
Pirenella insculpta	Terebralia palustris
Pirenella layardi	Terebralia sulcata

PLANAXIDAE

Planaxis lineolatus · Planaxis sulcatus
Planaxis punctorostratus

TRIPHORIDAE

Triphora aegle · Mastonia servaini
Triphora arafra · Mastonia squalida
Triphora candefacta · Cautotriphora pyramidalis
Triphora cingulata · Viriola cancellata
Triphora granulifera · Viriola corrugata
Triphora loyaltiensis · Viriola incisa
Triphora perlata · Viriola morychus
Triphora tricincta · Viriola senafirensis
Mastonia ducosensis minor · Iniforis distinguenda
Mastonia maenades · Iniforis jousseaumei
Mastonia monilifera · Iniforis undata
Mastonia rubra

EULIMIDAE

Eulima gentiluomiana · Balcis acuta
Eulima lactea · Balcis cumingi
Eulima muelleriae

STILIFERIDAE

Stilifer cumingianus · Stilifer thielei

JANTHINIDAE

Janthina globosa · Janthina janthina

EPITONIIDAE

Epitonium aculeatum · Epitonium irregulare
Epitonium adjunctum · Epitonium jomardi
Epitonium alatum · Epitonium lamellosum
Epitonium amica · Epitonium latedis junctum
Epitonium attenuatum · Epitonium replicatum
Epitonium crassilabrum · Epitonium robillardi
Epitonium fererussaci · Epitonium rubrolineatum
Epitonium gradilis · Epitonium scalare
Epitonium gravieri · Scaliola elata
Epitonium hyalinum

HIPPONICIDAE

Hipponix australis · Hipponix conicus

VANIKORIDAE

Vanikora distans

CALYPTRAEIDAE

Cheiles equestris

XENOPHORIDAE

Xenophora calculifera

FOSSARIDAE

Fossarus erythraoensis · Fossarus lamellosus

TRICHOTROPIDIDAE

Amathina tricarinata

STROMBIDAE

Strombus bulla · Strombus plicatus
Strombus debelensis · Strombus terebellatus
Strombus decorus · Strombus tricornis
Strombus dentatus · Strombus urceus
Strombus erythrinus · Terebellum terebellum
Strombus fasciatus · Lambis lambis
Strombus fusiformis · Lambis truncata sebae
Strombus gibberulus albus · Tibia insulaechorab
Strombus mutabilis

NATICIDAE

Natica areolata · Natica undulata
Natica maculosa · Polinices melanostoma
Natica mammilla (= Polinices m.) · Polinices tumidus
Natica marochiensis · Eunaticina mienisi
Natica onca · Sinum papilla
Natica powisiana

TRIVIIDAE = ERATOIDAE

Trivia exigua · Trivia pellucidula
Trivia oryza

OVULIDAE

Ovula marginata · Calpurnus verrucosus
 (= Pseudosimnia) · Pseudocypraea adamsonii
Calpurnus lacteus

CYPRAEIDAE

Cypraea arabica (= C. grayana) · Cypraea lynx
Cypraea asellus · Cypraea macandrewi
Cypraea bistrinotata · Cypraea microdon chrysalis
Cypraea camelopardalis · Cypraea erosa nebrites
Cypraea carneola · Cypraea nucleus
Cypraea caurica · Cypraea pantherina
Cypraea cicerula · Cypraea pulchra
Cypraea clandestina · Cypraea punctata
Cypraea cribraria · Cypraea staphylea
Cypraea erythraeensis · Cypraea talpa
Cypraea exusta · Cypraea teres
Cypraea felina · Cypraea teulerei
Cyrpaea fimbriata · Cypraea thomasi
Cypraea gangranosa · Cypraea turdus
Cypraea globusus · Cypraea chinensis variolaria
Cypraea gracilis notata · Cypraea vitellus
Cypraea helvola · Cypraea walkeri
Cypraea hirundo · Cypraea ziczac
Cypraea isabella · Monetaria annulus (= Cypraea)
Cypraea kieneri · Monetaria moneta (= Cypraea)
Cypraea lentiginosa

TONNIDAE

Tonna canaliculata · Tonna sulcosa
Tonna perdix · Malea pomum

FICIDAE

Ficus ficus · Ficus subintermedius

CASSIDIDAE

Phalium bisulcatum · Casmaria ponderosa unicolor
Phalium faurotis · Cassis cornuta
Casmaria erinaceus erinaceus

CYMATIIDAE

Cymatium clandestinum · Cymatium vespaceum
Cymatium exile · Charonia tritonis
Cymatium pileare · Apollon pusillus
Cymatium labiosum · Distorsio anus
Cymatium lotorium · Gutturinum trilineatum
Cymatium marerubrum · (= Cymatium)

COLUBRARIIDAE

Colubraria antiquata · Phyllocoma convoluta
Colubraria obscura

BURSIDAE

Bufonaria rana · Bursa spinosa
Bursa affinis granularis · Lampadopsis rhodostoma
Bursa bubo (= Tutufa) · (= Bursa)
Bursa bufonia · Tutufa rubeta
Bursa lamarcki

MURICIDAE

Murex scolopax · Pterynotus tripterus
Murex tribulus · Ocinebra contracta
Murex virgineus (= Chicoreus) · Thais bufo
Aspella anceps · Thais carinifera
Aspella producta · Thais hippocastanum
Chicoreus corrugatus · (= T. savignyi)
Chicoreus ramosus · Thais tuberosa
Phyllonotus jickelii · Nassa serta
 (= Naquetia) · Drupa fiscella
Favartia cyclostoma · Drupa lobata
Hexaplex turbinatus · Drupa morum
Homalocantha anatomica · Drupa ocbrostoma
Homalocantha dovpeledi · (=Mancinella mancinella)

MURICIDAE continued

Homalocantha digitatus
Haustellum haustellum
Naquetia annandalel
Vitularia miliaris
Purpura persica
Cronia martensi
Rapana rapiformis
Spinidrupa euracantha
Homalocantha scorpio
Pterynotus martinitana

Drupa ricinus
Drupa rubusidaeus
Drupella rugosa
Maculotriton serriale
Morula anaxeres
Morula granulata
Morula ocellata
Morula uva

CORALLIOPHILIDAE = MAGILIDAE

Rapa bulbosa
Mipus gyratus
Magilus antiquus
Coralliophila costularis

Coralliophila erosa
Coralliophila violacea
Quoyula madreporarum

BUCCINIDAE

Cantharus dorbignyi
Cantharus fumosus
Cantharus lanceola
Cantharus puncticulatus
Cantharus rubiginosus
Cantharus undosus

Engina alveolata
Engina mendicaria
Engina pulchra
Phos roseatus
Phos senticosus
Pisania ignea

COLUMBELLIDAE = PYREMIDAE

Columbella albina
Columbella albinodulosa
Columbella azora
Columbella conspersa
Columbella exilis
Columbella eximia
Columbella erythraeensis
Columbella flava
Columbella galaxias
Columbella mindoroensis

Columbella nomanensis
Columbella nympha
Columbella rustica
Columbella savignyi
Columbella terpsichore
Columbella testudinaria
  (= Pyrene)
Columbella tringa
Columbella troglodytes
Columbella varians

FASCIOLARIIDAE

Fasciolaria filamentosa
  (= Pleuroploca)
Fasciolaria trapezium
  (= Pleuroploca)
Latirus polygonus
Latirus turritus
Dolicholatirus lancea

Peristernia incarnata
Peristernia nassatula
Fusinus polygonoides
  (= Fusus)
Fusinus tuberculatus
Latirolagena smaragdula

NASSARIIDAE

Nassarius albescens
  gemmuliferus
Nassarius arcularia plicatus
Nassarius cinctellus
Nassarius concinnus
Nassarius crenulatus
Nassarius delicatus
Nassarius fissilabrus
Nassarius gemmulatus
Nassarius glans
Nassarius pauperus

Nassarius protrusidens
Nassarius thaumasius
Nassarius unicolor
Nassa lathraia
Nassa munda
Nassa obockensis
Nassa sporadica
Nassa steindachneri
Nassa stiphra
Nassa xesta

CANCELLARIIDAE

Scalptia scalata

MELONGENIDAE

Volema pyrum nodosa

HARPIDAE

Harpa minor = amouretta   Harpa ventricosa

VASIDAE

Vasum turbinellus

MARGINELLIDAE

Marginella monilis
Marginella pygmaea

Marginella savignyi
Marginella suezensis

OLIVIDAE

Ancilla castanea
Ancilla cinnamomea
Ancilla crassa
Ancilla lineolata
  (= A. acuminata)

Oliva elegans
Oliva inflata (= O. bulbosa)
Oliva picta

MITRIDAE

Mitra acutilirata
Mitra aurantia
Mitra aurora
Mitra bovei
Mitra chrysalis
Mitra cucumerina
Mitra fasciolaris
Mitra fissurata (= Scabricola)
Mitra fraga
Mitra imperialis
Mitra incompta
Mitra litterata
Mitra mitra
Mitra mucronata
Mitra nubila

Mitra paupercula
Mitra retusa
Mitra rueppelli
Mitra tabanula
Mitra typha
Mitra variegata
Cancilla carnicolor
Cancilla filaris
Neocancilla pretiosa
Scabricola desetangsii
Scabricola scabriuscula
  (= Neocancilla granatina)
Subancilla annulata
Subancilla flammea
  (= Mitra interlirata)

COSTELLARIIDAE (= VEXILLIDAE)

Vexillum amabile
Vexillum aureolata
Vexillum cadaverosum
Vexillum coronatum
Vexillum depexum
Vexillum fulvosulcatum
Vexillum infaustum
Vexillum kraussi
Vexillum leucozonias
Vexillum lubens
Vexillum lucidum
Vexillum macandrewi
Vexillum melongena
Vexillum osidiris

Vexillum pardalis
Vexillum rugosum
  intermediatum
Vexillum tusum
Vexillum unifasciale
Vexillum virgo
Costellaria casta
Costellaria deshayesi
  (= Vexillum)
Costellaria exasperata
  (= Vexillum)
Costellaria obeliscus
Pterygia crenulata

CONIDAE

Conus achatinus
  (= C. monachus)
Conus aculeiformis torensis
Conus acuminatus
Conus acutangulus
Conus arenatus
Conus auliscus
Conus capitaneus
Conus catus
Conus connectens
Conus coronatus
Conus cuvieri
Conus distans
Conus ebraeus
Conus emaciatus
Conus erythraeensis
Conus flavidus
Conus frigidus
Conus fulgetrum
Conus fumigatus
Conus geographus
Conus litoglyphus
Conus lividus

Conus magus
Conus maldivus
Conus miles
Conus miliaris
Conus musicus
Conus namocanus
Conus nussatella
Conus pennaceus
Conus planiliratus batheon
Conus quercinus
Conus rattus
Conus semivelatus
Conus sponsalis
Conus striatellus
Conus striatus
Conus taeniatus
Conus terebra
Conus tessulatus
Conus textile
Conus vexillum
Conus vicarius
Conus virgo
Conus vitulinus

TEREBRIDAE

Terebra affinis
Terebra albomarginata
Terebra amanda
Terebra areolata
Terebra argus
Terebra babylonia
Terebra cancellata
Terebra cerithina
Terebra columellaris
Terebra consobrina
Terebra crenulata
Terebra dimidiata

Terebra duplicata
Terebra evoluta
Terebra flavofasciata
Terebra insalli
Terebra lima
Terebra maculata
Terebra nebulosa
Terebra parkinsoni
Terebra textilis
Impages hectica
Hastula albula

TURRIDAE

Gemmula amabilis	Drillia flavidula
Xenoturris cingulifera	Carinapex minutissima
Turris amicta	Hemilienardia balteata
Turris violacea	Hemilienardia malletti
Tritonoturris cumingii	Lienardia mighelsi
Turricula catena	Kermia daedalea
Lophiotoma brevicaudata	Mangilia mica
Lophiotoma acuta jickeli	Mangilia rubida
Clavus acuminata	

PYRAMIDELLIDAE

Pyramidella acus	Pyramidella sulcata
Pyramidella mitralis	Pyramidella terebelloides
(= Otopleura)	Otopleura auriscati

ACTEONIDAE

Pupa solidula	Tornatina fusiformis
Pupa sulcata	Tornatina inconspicua
Acteon flammeus	Tornatina simplex
Acteon tornatilis	

RINGICULIDAE

Ringicula acuta

HYDATINIDAE

Hydatina amplustre	Hydatina physis

BULLIDAE

Bulla ampulla	Chelidonura flavolobata

ATYIDAE

Atys chelidon	Atys naucum
Atys cylindricus	Haminoea pemphix
Atys ehrenbergi	Dinia dentifera
Atys lithiensis	

RETUSIDAE

Retusa desgenettesi	Retuss fourieri

PHILINIDAE

Philine vaillanti	Lathophthalmus smaragdinus

SCAPHANDRIDAE

Cylichna girardi	Cylichna semisulcata
Cylichna mongei	

AGLAJIDAE

Aglaja cyanea

PLEUROBRANCHIDAE

Pleurobranchus forskali	Berthella oblonga
Berthella citrina	

PNEUMODERMATIDAE

Pneumoderma peronii

ELYSIIDAE

Elysia sp.	Elysia decorata

CAVOLINIIDAE

Cavolinia longirostris	Diacria quadridentata
Cavolinia uncinata	

APLYSIIDAE

Aplysia cornigera	Dolabella auricularis
Aplysia dactylomela	Dolabella gigas
Aplysia fasciata	Dolabella scapula
Aplysia oculifera	Dolabrofera dolabrifera
Aplysia parvula	Notarchus indicus
Paraplysia geographica	Notarchus savignyanus
Petalifera petalifera	

DISCODORIDIDAE

Discodoris concinna	Platydoris scabra
Sebadoris crosslandi	

CHROMODORIDIDAE

Chromodoris annulata	Casella obsoleta
Chromodoris fidelis	Casella rufomarginata
Chromodoris pallida	Glossodoris quadricolor
Chromodoris pantherina	Glossodoris runcinata
Chromodoris pulchella	Hexabranchus sanguineus
Casella atromarginata	Hypselodoris infucata
Casella cincta	Hypselodoris maridadalus

KENTRODORIDIDAE

Kentrodoris annuligera	Kentrodoris funebris

DENDRORIDIDAE

Dendrodoris cuprea	Dendrodoris pustulosa
Dendrodoris nigra	Dendrodoris rubra

PHYLLIDIIDAE

Phyllidia dautzenbergi	Phyllidia varicosa
Phyllidia arabica	Phyllidiella pustulosa
Phyllidia nobilis	

AEOLIDIIDAE

Aeolidiella orientalis	Baeolidia moebii

CUTHONIDAE

Catriona susa	Njurja netsica

FAVORINIDAE

Favorinus horridus	Pteraeolidia semperi
brevitentaculus	

FIMBRIIDAE = TETHYDIDAE

Melibe bucephala

TRITONIIDAE

Tritoniopsilla elegans

ELLOBIIDAE

Melampus lividus	Melampus flavus

SIPHONARIIDAE

Siphonaria kurracheensis

DENTALIIDAE

Dentalium anatorum	Dentalium lineolatum
Dentalium bisexangulatum	(= D. reevei)
	Dentalium longirostrum

ARCIDAE

Arca afra erythraea	Arca ventricosa
Arca antiquata (= Anadara)	Scapharca vellicata
Arca auriculata	Barbatia decussata
Arca avellana	Barbatia fusca
Arca clathrata	Barbatia helblingi
Arca ehrenbergi	Barbatia lacerata
Arca imbricata arabica	Barbatia lima
Arca natalensis	Barbatia nivea
Arca navicularis	Barbatia setigera
Arca plicata	Barbatia tenella
Arca uropygmelana (= Anadara)	

LIMOPSIDAE

Limopsis multistriata

GLYCIMERIDIDAE

Glycimeris arabicus	Glycimeris pectunculus
Glycimeris lividus	

MYTILIDAE

Modiolus arborescens	Crenella ehrenbergi
Modiolus arcuatulus	Septifer bilocularis
Modiolus auriculatus	Lithophaga hanleyana
Modiolus cinnamomeus	Lithophaga lessepsiana
Modiolus cumingianus	Lithophaga lima
Modiolus glaberrinus	Lithophaga malaccana
Modiolus ligneus	Lithophaga obesa

MYTILIDAE continued

Modiolus subsulcatus
Bracbydontes variabilis
Crenella adamsiana

Lithophaga robusta
Lithophaga teres
Botula cinnamomea

PINNIDAE

Pinna bicolor
Pinna muricata

Atrina vexillum
Streptopinna saccata

PTERIIDAE

Pteria aegyptiaca
Pteria crocata
Pteria macroptera
Pteria penguin
Pteria placunoides
Pteria spadicea

Pteria zebra
Electroma ala-corvi
Pinctada margaritifera
Pinctada vulgaris
 (= P. radiata)

ISOGNOMONIDAE

Isognomon australicum
Isognomon isognomon
Isognomon legumen

Isognomon nucleus
Crenatula picta

MALLEIDAE

Malleus regula
Vulsella attenuata
Vulsella rugosa

Vulsella spongiarum
Vulsella vulsella

PECTINIDAE

Pecten erythraeensis
Pecten lividus (= Chlamys)
Mirapecten rastellum
Gloripallium pallium
Chlamys irregularis
Chlamys lemniscata
Chlamys lucelentus
Chlamys sanguinolentus
 (= gloripallium)

Chlamys senatorius
Chlamys squamatus
Chlamys squamosus
Semipallium fulvico stata
Semipallium jousseaumei
Decatopecten plica
Pedum spondyloideum
Juxtamusium maldivensis

PLICATULIDAE

Plicatula australis

Plicatula plicata

SPONDYLIDAE

Spondylus aculeatus
Spondylus aurantius
Spondylus candidus
Spondylus castus
Spondylus ducalis

Spondylus gaederopus
Spondylus hystrix
Spondylus marisrubri
Spondylus regius

ANOMIIDAE

Anomia nobilis

LIMIDAE

Lima annulata
Lima fragilis

Lima lima

GRYPHAEIDAE

Hyotissa hyotis

Hyotissa numisma

OSTREIDAE

Ostrea crenulifera
 (= Alectryonella)
Ostrea forskali

Lopha cristagalli
Lopha folium
 (= Dendrostrea)

LUCINIDAE

Lucina dentifera
Lucina edentula
 (= Anodontia)
Lucina fieldingi
Lucina picta
Lucina semperiana
Codakia divergens (= Ctena)
Codakia tigerina

Loripes clausus
Loripes concinnus
Loripes erythraeus
Loripes fischerianus
Divaricella macandreae
Divaricella quadrisulcata
 (= D. ornata)

UNGULINIDAE

Diplodonte globosa
Diplodonta rotundata

Diplodonta tumida

CHAMIDAE

Chama asperella
Chama aspersa
Chama brassica
Chama cornucopia
 (= C. rueppelli)
Chama fragum

Chama fimbriata
Chama imbricata
Chama lazarus
Chama limbula
Chama pacifica
Chama reflexa

MONTACUTIDAE

Montaguia lamellifera

GALAEOMMATIDAE

Scintilla ovulina
Scintilla pisum

Scintilla variabilis
Galeomma mauritiana

CARDITIDAE

Cardita castanea
Cardita gubernaculum
 (= Beguina)
Cardita muricata

Cardita variegata
Venericardia akabana
Venericardia rufa

CRASSATELLIDAE

Eucrassatella jousseaumei

TRAPEZIIDAE

Trapezium bicarinatum
Trapezium oblongum

Trapezium sublaevigatum
Coralliophaga coralliophaga

CARDIIDAE

Cardium arabicum
Acanthocardia pseudolima
Trachycardium arenicola
Trachycardium pectiniforme
Trachycardium subrugosum
Papyridea australe
Papyridea papyracea
 (= Laevicardium)
Hemicardium auricula

Hemicardium fragum
Hemicardium fornicatum
Hemicardium nivale
Hemicardium retusum
Nemocardium exasperatum
Nemocardium lyratum
Laevicardium biradiatum
Laevicardium orbita
Laevicardium sueziense

TRIDACNIDAE

Tridacna crocea
Tridacna maxima

Tridacna squamosa

MACTRIDAE

Mactra achatina
Mactra liliacea
Mactra olorina

Standella solanderi
Lutraria oblonga

MESODESMATIDAE

Mesodesma striata
Atactodea glabrata

Ervilia purpurea
Ervilia scaliola

SOLENIDAE

Solen corneus
Solen lischkeanus

Solen truncatus

CULTELLIDAE

Cultellus cultellus
Siliqua polita

Solenocurtus australis
Solenocurtus coarctatus

TELLINIDAE

Angulus corbis
Quadrans gargadia
Tellina adamsi
Tellina arsinoensis
Tellina asperrina
Tellina bertini
Tellina crucigera
Tellina dubia
Tellina edentula
Tellina flacca
Tellina foliacea
Tellina inflata
Tellina isseli
Tellina lamellosa
Tellina nux
Tellina ovalis

Tellina pinguis
Tellina pristis
Tellina pulcherrima
Tellina rastellum
Tellina rhomboides
Tellina rugosa
Tellina sericata
Tellina speciosa
Tellina subpallida
Tellina sulcata
Tellina staurella
Tellina triradiata
Tellina truncata
Tellina valtonis
Tellina virgata
Tellina vulsella

TELLINIDAE continued

Tellina perna	*Cyclotellina scobinata*
Tellina petalina	*(= Scutarcopagia)*
Tellina pharaonis	*Quidnipagus palatum*

DONACIDAE

Donax abbreviatus	*Donax dohrnianus*
Donax clathratus	*Donax erythraeensis*

PSAMMOBIIDAE

Gari bicarinata	*Psammobia pulchella*
Gari granulifera	*Soletellina rubra*
Gari maculosa	*(= Hiatula ruppelliana)*
Gari weinkauffi	*Asaphis violascens*
Psammobia elegans	

SEMILIDAE

Semele striata	*Leptomya rostrata*
Abra lactea	

VENERIDAE

Venus lacerata	*Lioconcha castrensis*
Venus lamellaris	*Lioconcha picta*
Venus reticulata	*Dosinia alta*
(= Periglypta)	*Dosinia erythraea*
Venus verrucosa	*(= D. radiata)*
Circe arabica	*Dosinia hepatica*
Circe corrugata	*Dosinia histrio*
Circe crocea	*Dosinia pubescens*
Circe radula	*Clementia papyracea*
Circe scripta	*Chione hypopta*
Circe sulcata	*Marcia hiantina*
Circe undatina	*Tapes deshayesi*
Gefrarium divaricatum	*Tapes literatus*
Gefrarium pectinatum	*Tapes sulcarius*
Pitar affinis	*Notirus macrophyllus*
Pitar hebraea	*Paphia textile*
Amiantis philippinarum	*Timoclea barashi*
Sunetta effosa	*Timoclea costellifera*
Callista erycina	*Timoclea marica*
Callista florida	
(= Lepidocardia)	

PETRICOLIDAE

Petricola hemprichi	*Petricola lapicida*

MYIDAE

Cryptomya decurtata	*Sphenia rueppelli*

CORBULIDAE

Corbula erythraeensis	*Corbula sulcosa*
Corbula modesta	

GASTROCHAENIDAE

Gastrochaena cuneiformis	*Gastrochaena dentifera*
Gastrochaena cybium	*Spengleria mytiloides*
Gastrochaena deshayesi	

PHOLADIDAE

Pholas dactylus

LATERNULIDAE

Laternula anatina	*Laternula subrostrata*

CUSPIDARIIDAE

Cuspidaria brachyrhynchus	*Cuspidaria potti*
Cuspidaria dissociata	*Cuspidaria steindachneri*

LYONSIIDAE

Lyonsia intracta

CLAVAGELLIDAE

Clavagella adenensis	*Penicillus vaginifer*
	(= Brechites attrahens)

SEPIIDAE

Sepia pharaonis	*Sepia dollfusi*

LOLIGINIDAE

Sepioteuthis lessoniana

OMMASTREPHIDAE

Ommastrephus arabicus	*Symplectoteuthis oualensis*

ENOPLEUTHIDAE

Enoploteuthis dubia

OCTOPIDIDAE

Octopus aegina	*Octopus horridus*
Octopus cyaneus	*Octopus macropus*

8. BRYOZOANS

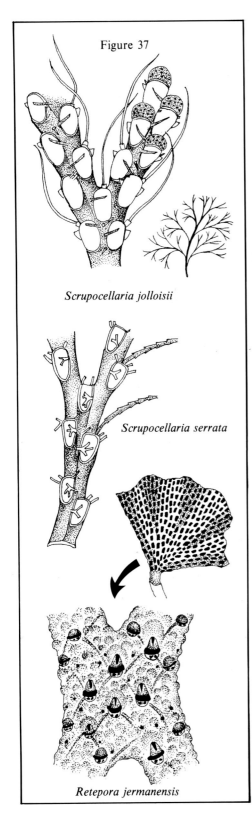

Figure 37

Scrupocellaria jolloisii

Scrupocellaria serrata

Retepora jermanensis

PHYLUM: BRYOZOA

Bryozoans are colonial, sessile animals in which individuals or zooids are usually less than half a millimeter in length. In most cases, the zooids are encased in a non-living envelope which contains an opening for the protrusion of a food-catching organ or lophophore. Whereas most bryozoans form flat encrusting colonies, some are plant-like and others form erect plates and are known as "lace-corals". Within a particular colony, in which zooids bud off from each other by asexual reproduction, there may be several kinds of zooids, each specialized for a particular function.

If one turns over a coral-slab the underside is likely to be heavily encrusted with a variety of organisms among which bryozoans are almost certain to be represented. Indeed, almost anywhere one looks closely at hard substrata in the Red Sea, bryozoans are likely to be discovered. They are among the first organisms to colonize newly exposed surfaces on coral-reefs. In so doing they play a significant role in cementing together coral fragments and thus consolidating the structure of the reefs. Identification of species is based upon microscopic examination of their zooids. Figure 37 illustrates a number of Red Sea bryozoans described in the text.

Scrupocellaria jolloisii Forms fan-shaped colonies which may be up to about 3cms in height. Zooecia have a small spine at each upper corner and a stout one which projects across the middle. It occurs on dead coral, frequently on the undersides of platy corals.

Scrupocellaria serrata is a related species which forms narrow branches. It occurs throughout the Red Sea and Gulf of Suez.

Bugula robusta forms bifurcating branches. Long, beaked avicularium grows from tubular bases close to proximal end of zoecium. Globular ovicell attached to side of zoecium.

Membranipora savartii forms mat-like encrustations spreading over the surface of the substrate which may consist of any hard surface, often on molluscan shells. The related species *Electra bellula* and *M. tuberculata* are usually found on *Sargassum* weed.

Celleporaria pigmentaria forms massive black colonies which frequently encrust corals.

Hippopetraliella magma forms broad tubes of about 10mm in diameter which anastomose irregularly, forming much larger colonies. It is particularly abundant on the sea-bed in the Bay of Agig where it is trawled up during shrimp fishing operations. The tubular branches may be inhabited by a species of *Chaetopterus* and by many other polychaetes and small crustaceans.

Stylopoma viride forms large greenish multi-layered encrustations on corals. The zooecia have a distinctive appearance consisting of ovate, raised forms with large pits in which there are one or two perforations at the base. Avicularia are large with long narrow mandibles. It is a common species in fairly shallow water where it encrusts stones, coral rubble, old mooring cables or other debris. In Suakin harbour, for instance, it covers the old telegraphic cables.

Robertsonidra argentea has granular zooecia in which a fairly large avicularium is diagonally placed.

Parasmittina egyptiaca is often present as an encrustation on the pearl oyster, *Pinctada margaritifera*. Zooecia are arranged in double rows with straight thick divisional walls between adjacent sections.

Retepora jermanensis seems to be a Red Sea endemic species and forms extremely delicate, lace-like colonies.

A list of Red Sea bryozoa is given in Table XVIII. It is likely that further research will reveal more species.

Membranipora savartii

Table XVIII
RED SEA BRYOZOA*

CHEILOSTOMATA ANASCA:

Aetea crosslandi
Acanthodesia limosa
Antropora granulifera, A. marginella
Nellia tenuis
Vibracellina viator
Canda arachnoides
Chaperina tropica
Steganoporella simplex
Thalamopella gothica indica
Thalamopella rozieri
Scrupocellaria diadema, S. jolloisii
S. longispinosa, S. obtecta, S. serrata
S. spatulata
Setosellina coronata
Smittipora cordiformis
Labioporella creulata
Beania cupulariensis, B. mirabilis
B. discodermiae
Bugula robusta
Membraniporella aragoi
Cribrilina radiata
Reginella floridana

CTENOSTOMATA:

Nolella papuensis
Buskia setigera

CHEILOSTOMATA ASCOPHORA IMPERFECTA:

Escharoides longirostris
Exechonella tuberculata, E. discoidea
Exechonella antillea
Tremogasterina robusta, T. spathulata
Celleporaria aperta, C. columnaris
C. granulosa, C. labelligera, C. pigmentaria
C. vermiformis, C. pilaefera
Cigliscula occlusa
Hippopetraliella magna
Mucropetraliella thenardii, M. echinata
M. philippinensis, M. robusta
Trematoecia turrita

CHEILOSTOMATA ASCOPHORA VERA:

Savignyella otophora
Margaretta gracilor, M. tenuis
Gigantipora fenestrata, G. pupa
Schizomavella australis, S. inclusa
S. sinapiformis
Stylopoma duboisii, S. viride
Scorpiondinipora bernardii
Cribellopora trichotoma
Calyptotheca wasinensis, C. nigra
C. sudanensis, C. capensis, C. subimmersa
C. thorneleyae
Emballotheca harmeri
Hippopodina feegeensis
Hippaliosina acutirostris
Schismopora redoutei
Watersipora subovoidea
Cosciniopsis lonchaea, C. globosa
Parasmittina egyptiaca, P. glomerata
P. parasevalii, P. raigii, P. tropica
P. protecta, P. signata, P. tropica
Codonellina montferrandii
Celleporina costazii
Robertsonidra argentea
Drepanophora incisor, D. longiscula, D. birbira
Cleidoclasma laterale, C. porcellanum, C. protrusum
Rhynchozoon incrassatum, R. larreyi, R. tubulosum
Hippoporella multidentata, H. spinigea
Thorneyla ceylonica
Retepora jermanensis
Reteporellina denticulata
Sertella praetenuis
Triphyllozoon hirsutum
Anoteropora latirostris
Smittina nitidissima
Smittoidea levis
Chorizopora brongniartii
Trypostega venusta
Microporella orientalis
Escharina pesanseris
Arthropoma cecilii, A. circinatum
Schizobrachiella convergeus

* after Powell, 1969 and Dumont, 1981.

9. ECHINODERMS

PHYLUM: ECHINODERMATA

The Echinoderms include sea-stars, sea-urchins, feather stars, sea cucumbers and brittle stars. The name means "spiny-skinned" and for many species this is an apt description, but not all members are so armed. One fairly general feature is their five-rayed symmetry which may be quite conspicuously displayed or else only discernible in the skeletal structure. Echinoderms all have a skelcton, underlying the skin, which is formed by calcareous plates or ossicles which are sometimes fused and in other forms loosely distributed.

An abbreviated classification of the phylum (which deals only with those forms discussed in this chapter) is provided in Table XIX.

Table XIX			
CLASSIFICATION OF ECHINODERMATA			
PHYLUM	CLASS	SUB CLASS	ORDER
ECHINODERMATA	CRINOIDEA		
	STELLEROIDEA	ASTEROIDEA	
		OPHIUROIDEA	
	ECHINOIDEA		CIDAROIDA
			DIADEMATACEA
			ECHINACEA
			GNATHOSTOMATA
	HOLOTHUROIDEA		

Walking Crinoid

Class: Crinoidea: Feather Stars

Feather-stars which live in shallow water are best observed by night-diving in the Red Sea when they can be seen clinging to the reef edge with their feathery arms extended to form broad fans which sieve the plankton. The basic features of a generalized feather-star are illustrated in figure 38.

There is considerable diversity among shallow water feather-stars in the Red Sea with twelve or thirteen species regularly occurring within the depth range of 1 to 25m (Table XX). At times they can be quite densely aggregated with around 50 individuals per m² as normal, and up to 100/m² having been recorded. Such aggregations are important consumers of plankton and their significance in this regard has sometimes been overlooked. Many feather-stars play host to a variety of commensals which may, in turn, show a high degree of specificity with regard to their selection of host. The most notable feather-star in this regard is probably *Heterometra savignii* from which 18 species of commensals have been collected.

Table XX

RED SEA SHALLOW WATER FEATHER STARS

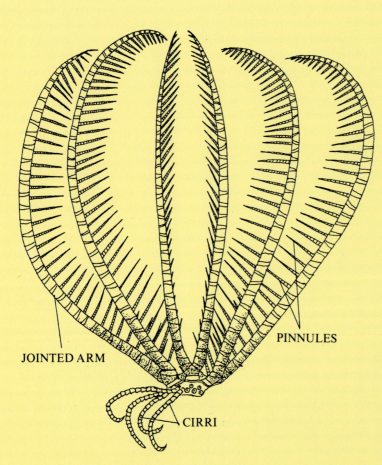

Figure 38: *Generalized Feather-star*

COMASTERIDAE
Capillaster multiradiatus
Comissia hartmeyeri

HIMEROMETRIDAE
Heterometra atra
H. savignii

MARIAMETRIDAE
Lamprometra klunzingeri
Stephanometra indica
S. spicata

COLOBOMETRIDAE
Decametra chadwicki
D. mollis
Oligometra serripinna

TROPIOMETRIDAE
Tropiometra carinata

ANTEDONIDAE
Dorometra aegyptica
Antedon parviflora

FAMILY: COMASTERIDAE

In this family the proximal pinnules are very flexible and some of the terminal segments are modified to form a comb. The mouth is near the edge of the disc and the anal tube is approximately central. Three species are present in the Red Sea: *Capillaster multiradiatus; Comissia hartmeyeri; Comaster distinctus.*

The first two may be separated, since *Comissia* is a smaller species with only ten arms (generally less than 40cms long), while *Capillaster multiradiatus* has up to forty or more arms but usually between 15 and 25cms in length.

C. multiradiatus and *C. hartmeyeri* occur in shallow water, with *Capillaster multiradiatus* generally mixed with populations of *Heterometra savignii* and *Lamprometra klunzingeri*, while *Comissia hartmeyeri* is often found in crevices along eroded seaward fringes of coral tables at one or two metres depth. They have quite different behaviour from *C. multiradiatus* or the other shallow water species mentioned above, since they are never found exposed but remain underwater, clinging to crevices and extending their arms into the current to capture plankton. If removed from their protected crannies they are relatively clumsy climbers and are vulnerable to being washed off the coral by wave surge. *Comaster distinctus* occurs at around 25m deep.

Feather-star: *Oligometra serripinna*

FAMILY: HIMEROMETRIDAE

This family includes the shallow water crinoids *Heterometra savignii* and *H. atra.* The former species usually has twenty arms and these have long spines towards their ends while *H. atra* has only fourteen arms. The latter form is a rare species and very few specimens have been studied. *H. savignii,* on the other hand, is one of the most conspicuous crinoids found on Red Sea coral reefs, where it occurs from the shallowest zone to around 12m depth. It is an important constituent of the group of feather-stars which are associated with *Millepora dichotoma* along the fore-reef, where they often occur in large numbers and may be observed feeding at night-time. In the lower half of its depth range (6m to 12m) it may be seen with its arms extended and feeding during daytime.

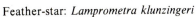

Feather-star: *Lamprometra klunzingeri*

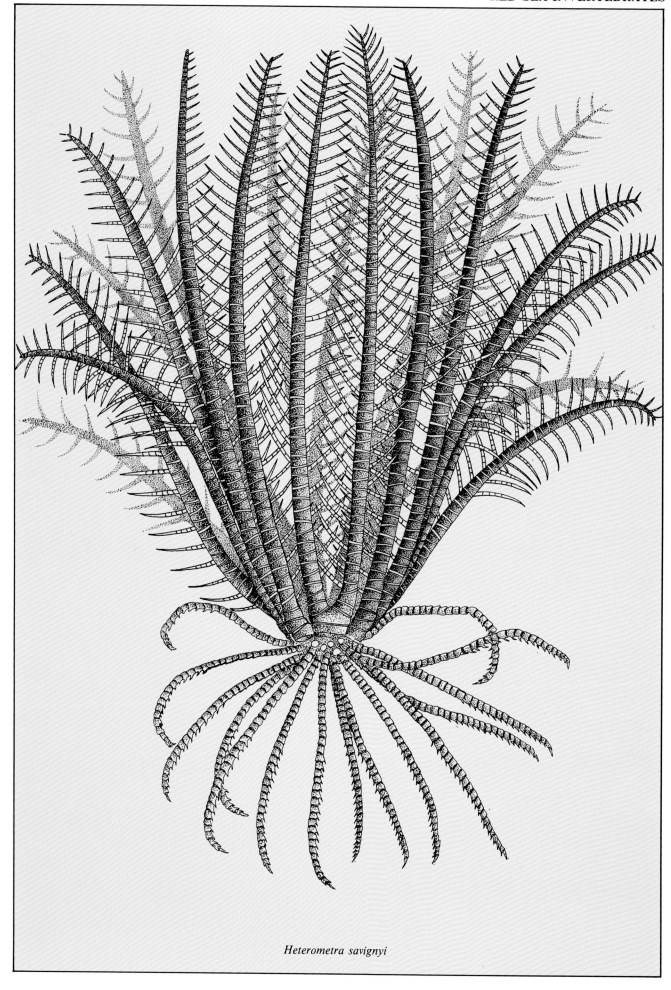

Heterometra savignyi

Lamprometra klunzingeri at night

FAMILY: MARIAMETRIDAE

All members of this family have more than ten arms; *Lamprometra klunzingeri*, *Stephanometra indica* and *S. spicata* are represented.

L. klunzingeri is frequently the dominant shallow water crinoid along *Millepora* dominated reef-faces (up to seventy per m²) and may even be found in crevices at low levels of the intertidal. It frequently hides among branches of *M. dichotoma* and rarely occurs more than about 2m deep. It has a distinct diurnal rhythm. At about an hour before sunset the first individuals may be seen to emerge from their daytime hiding places and using their curved arms they literally crawl to the upper branches of their coral refuges, finally taking up position by gripping with their cirri at the tips of the *Millepora* branches. Once they are firmly attached they spread their arms so that they form a plankton collecting sieve. An hour before sunrise they return to their hiding places.

Like all the other crinoids, *L. klunzingeri* is a host for several other parasitic or commensal creatures including the copepods *Pseudoanthessius major*, *P. minor*, *Collorchus uncinatus* and *Kelleria gradata*, the Palaemonidae shrimps *Pontoniopsis comanthi*, *Periclimenes tenuis* and *P. djiboutensis;* the brachyuran crab *Ceratocarcinus spinosus;* hermit crab *Galathea elegans;* and bristle worms (Myzostomida) which are especially adapted to life on crinoids, and are here represented by *Myzostomum crosslandi* and several other species of this genus. Mention should also be made of endoparasitic myzostomids, such as *Notopharyngoides ijimani* found on *Capillaster multiradiatus;* the aphroditid worm *Scelisetosus longicirrus*, which occurs on most crinoids; parasitic molluscs such as *Melanella sp.* and *Mucronalia capillastericola;* and finally, of the clingfish *Lepidichthys lineatus* which cling with their sucking discs to the dorsal sides of the host's arms.

FAMILY: COLOBOMETRIDAE

Two crinoids which occur in relatively deep water, lower down the reef-face (below 25m) are *Decametra chadwicki* and *Oligometra serripinna,* both members of the family Colobometridae. Towards the upper level of their depth range these two may be mixed with *Decametra arabica* and with the previously mentioned Comasterid feather-star, *Comaster distinctus,* but below 30m the only dominant species are *D. chadwicki* and *O. serripinna,* which cling with their hook-like cirri to gorgonians, black corals or sponges. Unlike the shallow water crinoids, these do not exhibit any marked diurnal behavioural rhythms but frequently remain fixed to the one place, in an arms extended feeding posture, for months at a time without changing their location. A third species of this family, *Decametra mollis,* occurs in the Gulf of Suez.

FAMILY: TROPIOMETRIDAE

The reef-face from about 5m to 20m has an intermediate selection of crinoids, including some which are generally found in the shallows and others which are more characteristic of deeper water. Included in this group is *Tropiometra carinata* which is usually hidden in holes in areas of coral rock. It is also found on reef-flats of the Gulf of Aqaba. It has ten arms which have conspicuously straight and stiff pinnules.

FAMILY: ANTEDONIDAE

Caves and crevices in fairly deep water are frequently colonized by soft corals such as *Dendronephthya* and by various sponges. Small ten armed crinoids of the Antedonidae may be associated with these. A species usually found hanging from the roofs of these caves, frequently in large numbers, is *Antedon parviflora.* This has one interesting facet to its behaviour which is not found among the other crinoids which have been mentioned. If a diver disturbs a group of these feather-stars they swim away, using graceful undulating strokes of their arms to propel themselves. Another member of this family found in the same microhabitat is the small feather-star *Dorometra aegyptica.*

Crinoid: *Antedon parviflora* on soft coral: *Dendronephthya* sp.

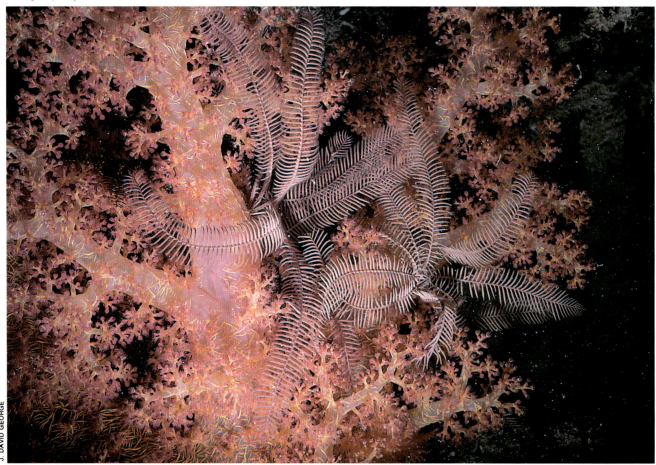

Class: Stelleroidea

Subclass: Asteroidea: Sea-stars

Starfish associated with coral-reefs show a range of adaptations to their environment. While most species graze on algae or browse on encrusting fauna, *Acanthaster planci* (the Crown of Thorns starfish) and *Culcita coriacea* (Cushion-star), predate upon live corals. Whereas *A. planci* rarely eats anything else, *Culcita* has a much more varied diet and does not depend upon corals as its food source. Most starfish associated with coral-reefs are extremely vulnerable to predators and they tend therefore to remain well hidden during daytime and to emerge at dusk.

In the case of *A. planci* this diurnal rhythm is particularly well established and this species has the additional advantage of a spiny armament which protects it against all but the most persistent of predators.

A generalized seastar is shown in figure 39 and a list of Red Sea species is provided in Table XXI.

Table XXI

RED SEA STARFISH

LUIDIIDAE

Luidia maculata, L. prionota, L. savignyi

ASTROPECTINIDAE

*Astropecten bonnieri, A. hemprichi,
A. monacanthus, A. orsinii, A. polyacanthus*

GONIASTERIDAE

*Monachaster umbonatus, Omgaster capella,
Stellaster equestris, Stellasteropsis fouadi*

OREASTERIDAE

*Choriaster granulatus, Culcita coriacea
Pentaceraster mammillatus,
P. tuberculatus*

OPHIDIASTERIDAE

*Fromia ghardaqana, Gomophia egyptiaca
Leiaster leachi, L. coriaceus,
Linckia multiflora, Ophidiaster hemprichi*

ASTEROPIDAE

Asteropsis carinifera

ASTERINIDAE

Asterina burtoni

ACANTHASTERIDAE

Acanthaster planci

PTERASTERIDAE

Euretaster cribosus

MITHRODIIDAE

Mithrodia clavigera

ECHINASTERIDAE

Echinaster callosus, E. purpureus

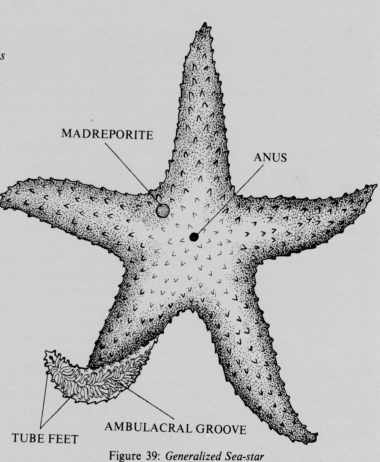

Figure 39: *Generalized Sea-star*

Pentaceraster mammillatus

Gomophia egyptiaca

Luidia maculata

FAMILY: LUIDIIDAE

These lack a terminal disc and have tube-feet which taper to a rounded or conical knob. Large skeleton plates define the edge of the body.

Luidia maculata (see drawing) has seven to nine long arms which lack prominent dorsal spines.

Luidia savignyi usually has seven arms which bear pedicellariae and spines on their dorsal surfaces.

FAMILY: ASTROPECTINIDAE: Comb Stars

These have a central disc and five arms which are fringed with spines rendering a comb-like appearance. Tube-feet lack suckers, and related to this is the fact that they are found in sandy areas where they tend to burrow just below the surface. They are carnivorous nocturnal predators, feeding mainly on invertebrates in the sand. Represented species are listed in Table XXI.

FAMILY: GONIASTERIDAE

This family includes a rather wide variety of forms from species with long, narrow well defined arms to those which are more like a pentagon with no discernible arms. A feature which links the family is the presence of large marginal plates delineating the outer edges of the starfish which usually has a flat upper surface. The junctions between arms are usually well-rounded rather than angular. Represented species are listed in Table XXI.

Linckia multiflora

FAMILY: OREASTERIDAE: Pin-cushion Starfish

These have large, convex discs with short arms which may merge completely with the disc, giving the animal a pentagonal shape. Large tubercles or spines are often present on the dorsal side which may have a thick skin covering. Some species such as *Culcita coriacea* feed on coral polyps in the same way as its more notorious relative *Acanthaster planci*.

Choriaster granulosus is completely covered with smooth, thick pink or orange skin and lacks any protruberances except for adambulacral spines bordering the furrow. It occurs in the Gulf of Aqaba and probably also in the main basin of the Red Sea. Represented species are listed in Table XXI.

Left: *Choriaster granulosus*
Below: *Gomophia egyptiaca*

F. JACK JACKSON

Gomophia egyptiaca

Below: *Asterina burtoni*
Opposite: *Fromia* sp.

HORST MOOSLEITNER

FAMILY: OPHIDIASTERIDAE

These starfish usually have small central discs and five long arms which are often more or less cylindrical in cross-section. The family includes three of the most frequently encountered Red Sea starfish, i.e. *Fromia ghardaqana, Gomophia egyptiaca* and *Linkia multiflora.*

FAMILY: ASTEROPIDAE

This family is represented by a single species, *Asteropsis carinifera* which is a drab, dark green-brown five armed sea-star with the outer edge fringed by blunt spines. It has a broad, flat underside and arms which are triangular in cross section with a median keel bearing spaced, conical spines similar to those fringing the margin. The remainder of the surface is smooth.

FAMILY: ASTERINIDAE

These have overlapping plates which resemble scales. Many of the plates have spines. The single Red Sea species, *Asterina burtoni* is a small sea-star which superficially resembles a miniature pin-cushion starfish in that it possesses five very short, rounded arms and its upper surface is markedly convex.

A. burtoni is one of the few species to have migrated through the Suez Canal and to have populated the Eastern Mediterranean. In the Red Sea there is another closely related form which has more than five arms. It has been named *Asterina wega* but is probably a variety of *A. burtoni.*

Above: *Ophidiaster hemprichi*
Right: *Fromia* sp.
Opposite: *Acanthaster planci*

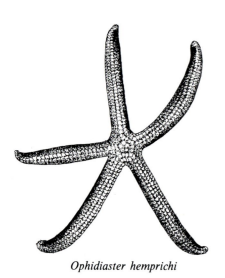

Ophidiaster hemprichi

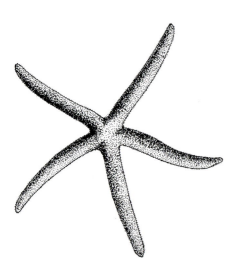

Echinaster purpureus

FAMILY: ACANTHASTERIDAE: Crown of Thorns Starfish

Many readers will be familiar with the spiny Crown of Thorns starfish, *Acanthaster planci*. This starfish has perfected the art of eating corals, a feat which it achieves by everting its stomach and spreading it over the surface of the coral, so that digestive enzymes secreted by the stomach come in direct contact with live coral tissues. Thus, as in many starfish, digestion commences *before* the food has been taken into the body through the mouth. Unlike many other coral predators, *A. planci* does not damage the coral skeleton but simply removes the live coral tissues. In fact it makes such a clean job of this, that one can easily identify the presence of the Crown of Thorns starfish by observing the blanched white feeding scars which remain after a night's feeding on coral. During daytime, *A. planci* usually remains in hiding and for a creature of its size, it has a remarkable ability to conceal itself in narrow shaded crevices. Aggregations occur naturally and may be associated with breeding or with feeding activity (or both). Individuals which form part of these aggregations frequently make less effort to hide during daytime and may simply crawl into the shade, under a coral table. Population explosions have been noted in many parts of the Indo-Pacific and Red Sea since the early 1960's. While Man has been implicated as a causative agent of these, it is this author's view that these have been naturally occurring and probably cyclical phenomena which have previously passed unnoticed, at least by the world's scientific community, if not by the corals themselves and their more intimate associates!

FAMILY: PTERASTERIDAE

Represented by a single species: *Euretaster cribosus* which has five short, blunt arms with aboral spines supporting a membrane covering the upper surface.

FAMILY: MITHRODIIDAE

These have an aboral skeleton which is more or less concealed by granules or scale-like tubercles which also cover the often conspicuously large spines. The represented species: *Mithrodia clavigera* is not uncommon at moderate depths. It has a mottled colour pattern and five reticulated cylindrical arms which are slightly constricted at the base and liable to break off. There are prominent ambulacral spines.

FAMILY: ECHINASTERIDAE

There are two species from this family in the Red Sea: *Echinaster callosus* and *E. purpureus*. These have five long, narrow arms with a very narrow ambulacral groove. Short spines cover the upper arm surface.

Basket star: *Astroba nuda*

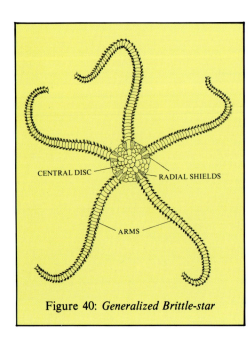

Figure 40: *Generalized Brittle-star*

Class: Stelleroidea

Subclass: Ophiuroidea: Brittle Stars

Brittle stars are among the less conspicuous echinoderms inhabiting coral-reefs. Most remain hidden in crevices or under stones and are only observed as a result of close inspection of the substratum. Nevertheless, there are around fifty species so far recorded from the Red Sea and these show a wide variety of adaptations to their life among corals. Virtually all shallow-water species are nocturnal and they are generally quite well hidden in daytime. They are relatively defenseless against predators so they tend to rely on fairly rapid escape responses or on cryptic camouflage to avoid being eaten. If they are attacked they may break off one or two arms in order to escape from the predator, but this is only a temporary inconvenience since they are able to regenerate lost arms. The arms themselves are sharply delineated from the disc and are segmented and flexible. The grooves which run along the oral sides of arms, unlike those of starfish, are covered over and the equivalent of their tube-feet are known as tentacles and are used in feeding rather than walking.

Brittle stars show a range of feeding habits from detrital feeders, through predatory forms of planktivorous species. Among the latter are the basket-stars whose branching arms are extended at night-time to form expansive fan-shaped plankton traps.

A generalized brittle star is shown in figure 40.

Table XXII

OPHIUROIDEA: BRITTLE STARS

OPHIOMYXIDAE
Lophiomyxa australis

GORGONOCEPHALIDAE
Astroba clavata, A. nuda

AMPHIURIDAE
Amphilycus scripta
Amphiodia microplax
Amphioplus personatus, A. timsae
A. hastatus, A. integer, A. laevis
Amphipholus squamata
Amphiura dejectoides
Dougaloplus echinatus

OPHIACTIDAE
Ophiactis carnea, O. hexacantha
O. parva, O. savignyi

OPHIOTRICHIDAE
Macrophiothrix demessa
M. galateae, M. hirsuta
Ophiomaza cacaotica
Ophiopsammium semperi
Ophiothela danae
Ophiothrix proteus, O. purpurea
O. propinqua, O. savignyi

OPHIOCOMIDAE
Ophiocoma erinaceus
O. macroplaca, O. pica,
O. pusilla, O. scolopendrina,
O. valenciae,
Ophiocomella sexradia
Ophiomastix variabilis
Ophiosila pantherina

OPHIONEREIDAE
Ophionereis dubia,
O. porrecta
O. variegata

OPHIODERMATIDAE
Ophiarchna incrassata
Ophioconis cupida
Ophiopeza fallax,
O. spinosa

OPHIURIDAE
Ophiolepis cincta,
O. superba
Ophiura kinbergi
Ophiocirce mabahithae

Macrophiothrix demesa

FAMILY: GORGONOCEPHALIDAE: Basket-stars

When one encounters living basket-stars during a night-dive they are most impressive creatures with their branching arms extended to form large fans. In this feeding posture they superficially resemble large feather-stars rather than their fellow brittle-stars. As soon as a torch beam is focussed on them they begin to curl up and, if the light is held in position, the previously delicate fan-shaped network of arm branches turns into a jumbled mass of arms which is more reminiscent of a tangle of old rope than the magnificent creature which one first noticed in its feeding posture at the reef-edge. A close look at such basket-stars will reveal that their branched arms bear rows of hooks across their upper surfaces.

Astroba nuda has a diameter up to almost a metre and its highly branched arms end in tendril-like tips. During daytime it hides in crevices but at night it emerges to take up position, usually on the same coral head as on previous nights, whereupon it spreads its arms into an open mesh basket and commences feeding on plankton. The related species *Astroba clavata* is also present in the Red Sea.

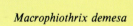
Astroba nuda

HORST MOOSLEITNER

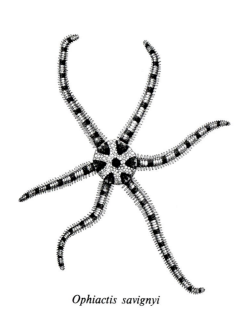

Ophiactis savignyi

FAMILY: AMPHIURIDAE

A diagnostic characteristic of this family is the presence of a pair of papillae at the apex of the jaw, below the lowest tooth. The disc may have distinct scales some of which can bear spines. Represented species are listed in Table XXII.

They are generally quite inconspicuous brittle-stars living buried in the sand or hidden among coral rubble. Some live in association with sponges and algae. Both *A. squamata* and *A. laevis* have very widespread distribution.

FAMILY: OPHIURIDAE: Ophiurids

In ophiurids the disc is covered with scales or plates rather than granulations. The genus *Ophiolepis* has a symmetrical arrangement of supplementary arm plates.

Ophiolepis cincta is a dull coloured species with a disc diameter rarely exceeding 15mm, while *O. superba* has a bold, pentaradiate disc pattern and purple arm bands on light brown. A third species, *Ophiura kinbergi* is also present. They live under stones or coral fragments during daytime and emerge to feed at night.

FAMILY: OPHIACTIDAE: Ophiactids

In these the teeth are broad and square-tipped but with only a single papilla (or reduced tooth) if any, at the superficial end of the column of teeth. The distal oral papillae may number one or two (occasionally three) and they are well spaced from the apex of the jaw. There is a single, fairly large, rounded tentacle scale. Most ophiactids are relatively small and are found living with sponges or among coral rubble.

Ophiactis savignyi is primarily epizoic in sponges but also hides in crevices on the reef. It usually has six arms but since it may multiply by fission, there are individuals with as few as two arms which have not completed regeneration of new arms following fission. Despite its small size, its conspicuous colour pattern of green with darker markings and large bare radial shields render it easily recognized. Other species are listed in Table XXII.

Ophiuroid: *Ophiocoma scolopendrina* feeding on surface film of water

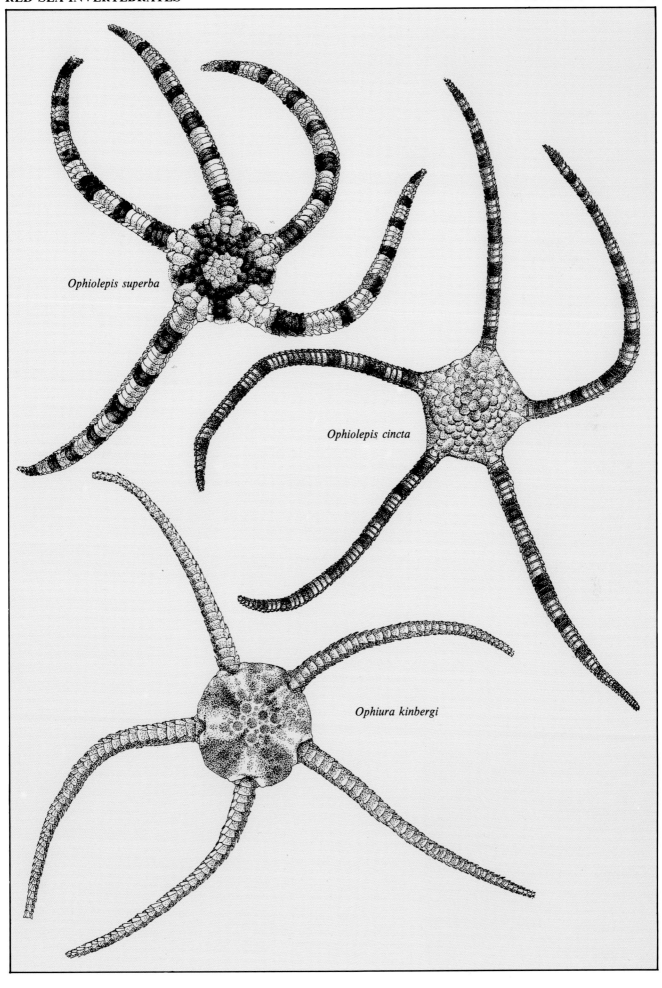

Ophiolepis superba

Ophiolepis cincta

Ophiura kinbergi

Ophiocoma scolopendrina

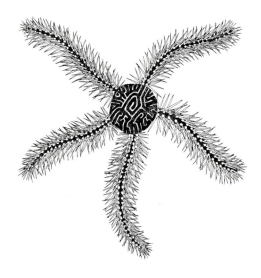

Ophiocoma pica

Ophiothrix purpurea
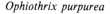

FAMILY: OPHIOTRICHIDAE: Ophiotrichids

This is a large family of brittle-stars with at least ten species present in the Red Sea. A diagnostic feature is the presence of a cluster of dental papillae at the apex of each jaw. There are no oral papillae. The disc may be covered with spines and the arms usually have prominent spines. Table XXII gives the represented species.

Ophiotrichid brittle-stars occupy a range of habitats associated with the reefs. Many live in crevices among coral rubble while others may live in association with sponges and gorgonians or with coralline algae. *Ophiothrix purpura,* which has a patterned disc and a purple line along the top of each arm, is a particularly common brittle-star in shaded coral caves and tunnels in the upper two metres or so of the reef-front. *Ophiothrix propinqua* prefers sand pockets on the reef-flat or sheltered crevices in shallow water near reef-edge.

FAMILY: OPHIOCOMIDAE: Ophiocomids

These are generally quite large brittle-stars in which the disc diameter can exceed two centimetres. Most species have a clump of dental papillae at each jaw apex (but in several species small individuals may have only a few). Oral papillae are present. Arms usually have prominent spines. In *Ophiocoma* the disc scales are more or less concealed by granules, and tentacle scales (usually two in number) are ovate. In *Ophiomastix* small spines are present on the disc. Most of these occur in association with hard substrates in coral crevices or in small caves or tunnel systems at the reef-edge. On the reef-flat they may be found under coral boulders.

Ophiocoma erinaceus is a black ophiuroid which lives sublittorally among live coral branches and among interstices of the stinging hydrozoan *Millepora. O. pica* has a similar habitat and is also dark but with a beautiful pattern of radiating gold lines on the disc.

Ophiocoma scolopendrina is always lighter underneath than on its upper surface and is the most abundant brittle-star in intertidal areas. It is a territorial species which protects its particular cranny or crevice from intrusion by other brittle stars. It feeds on organic detritus on the surface water film. *O. valenciae* has a similar habitat among loose rocks in shallow water. *O. porrecta* lives among rocks and is also found on coral sand.

FAMILY: OPHIOMYXIDAE

Disc and short arms are covered by a thick skin. Dorsal arm plates rudimentary and often fragmented. Represented by *Ophiomyxa australis* which is present in the Gulf of Aqaba but not so far recorded from the main basin of the Red Sea (see Table XXII).

FAMILY: OPHIONEREIDAE: Ophionerids

These have three upper arm plates in each arm segment and arms with prominent, laterally projecting spines. There is a single large tentacle scale on the lower surface of each arm segment. The arms are inserted below the disc which is usually covered with fine scales. See Table XXII for represented species.

FAMILY: OPHIODERMATIDAE: Ophiodermatids

These have arms which are fused with the sides of the densely granulated disc. Arm spines in the large, green, brittle-star *Ophiarchna incrassata* are laterally projecting and prominent.

Ophioconis cupida is a smaller brittle-star with much less prominent arm spines, while in *Ophiopeza fallax* reduced spines lie along the arms and are barely visible.

Ophiopeza spinosa is brownish in colour with eleven or twelve arm spines. It is recorded from the Gulf of Aqaba but not from the main basin of the Red Sea. Ophiodermatids are found on sand or coralline rubble under coral boulders or in shallow pools on the reef-flat.

Echinoid: *Microcyphus rousseaui*

Class: Echinoidea: Sea Urchins

Sea-urchins are well-known to those who dive or swim among Red Sea coral-reefs. Around fifty species have been recorded from the region but of these approximately twelve may be regarded as reasonably common. Sea-urchins are the only class of Echinoderms in which the skeletal plates are interlocked into a rigid theca or test. The more primitive urchins have regular, more or less spherical tests whereas more advanced urchins have irregular ones in which secondary bilateral symmetry may be imposed on the basic radial pattern, so that the urchin has a distinct anterior and posterior, along with a directional mode of movement.

The radial rows of interlocking skeletal plates are arranged in double columns consisting of ambulacral ones, which are distinguished by pores through which tube-feet pass, and interambulacral plates which lack these perforations. The remarkable arrangement of skeletal elements around the mouth can only be observed if an urchin is dissected.

A generalized sea urchin is shown in figure 41 and species found in the Red Sea are listed in Table XXIII.

There is no doubt that urchins play a major role in shallow reef ecology. Nearly all species are nocturnal and remain hidden during daytime. When they emerge at dusk their night-time feeding activities replace the intensive grazing and browsing over reef-surfaces carried out during daytime by many reef-fishes. If it were not for such grazing, rock surfaces would soon become coated by algae and, in more sheltered locations, clogged by sediment. Scraping of coral-rock surfaces by urchins is a major source of bio-erosion on Red Sea reefs and in turn a creator of "coral-sand". Freshly scraped surfaces are often favoured by settling invertebrate larvae such as those of coral planulae or serpulid trochophores. Without the creation of such surfaces many reef organisms would fail to locate suitable surfaces for their settlement.

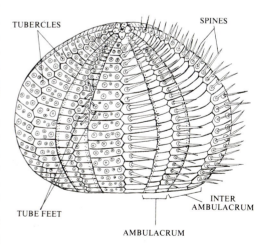

TUBERCLES   SPINES

TUBE FEET   INTER AMBULACRUM

AMBULACRUM

Figure 41: *Generalized Sea Urchin*

Table XXIII
RED SEA ECHINOIDS

CIDARIDAE
Eucidaris metularia
Phyllacanthus imperialis
Prionocidaris baculosa

ECHINOTHURIDAE
Asthenosoma varium

DIADEMATIDAE
Chaetodiadema granulatum
Diadema setosum
Echinothrix calamaris, E. diadema

TEMNOPLEURIDAE
Microcyphus rousseaui
Salmaciella dussumieri
Salmacis bicolor
Temnopleurus toreumaticus, T. scillae

TOXOPNEUSTIDAE
Nudechinus gravieri, N. scotiopremnus
Tripneustes gratilla

PARASALENIIDAE
Parasalenia poehli

ECHINOMETRIDAE
Echinometra mathaei
Echinostrephus molaris
Heterocentrotus mammillatus
H. trigonarius

CLYPEASTERIDAE
Clypeaster amplificatus, C. fervens
C. humilis, C. rarispinus, C. reticulatus

FIBULARIIDAE
Echinocyamus crispus, E. elegans
Fibularia ovulum, F. volva

LAGANIDAE
Laganum depressum, L. joubini

SCUTELLIDAE
Echinodiscus auritus
E. bisperforatus

ECHINOLAMPADIDAE
Echinolampas alexandri
Palaeostoma mirabile

LOVENIIDAE
Lovenia elongata

SPATANGIDAE
Maretia planulata

SCHIZASTERIDAE
Diploporaster savignyi
Moira stygia
Schizaster lacunosus
Paraster gibberulus

BRISSIDAE
Brissopsis luzonica
Brissus latecarinatus
Metalia spatangus, M. sternalis
Pericosmus akabanus

FAMILY: CIDARIDAE: CIDARIDS

Cidarids are regular echinoids which have large, well-spaced primary spines which lose their skin covering and thus allow settlement of encrusting organisms. These spines are ringed by small, more or less spatulate secondary spines. The test is composed of rigid plates (except for the peristome), and ambulacral colums are narrower than inter-ambulacral ones. Perforations for tube-feet in the ambulacral plates are in a single vertical series.

Represented species are: *Eucidaris metularia*, *Phyllacanthus imperialis* and *Prionocidaris baculosa*.

P. baculosa is relatively common on shallow, sheltered reef areas. Its primary spines are basically spotted. In the Gulf of Suez they migrate into deeper water during spring (around April-May) and return to the shallows in December-January. It is an omnivorous species feeding on a variety of algal and associated material, including fragments of live coral, molluscs and crustaceans. In the Northern Red Sea, *P. baculosa* spawns in July and August.

P. imperialis occurs in somewhat more exposed areas than *P. baculosa* and is often seen lodged in coralline crevices at the reef-edge or the micro-atoll zone behind the reef crest.

E. metularia is usually found at the reef-edge or on the fore-reef slope. Its diameter is about 25mm and the upper surface of its test normally displays a beautiful pattern which contrasts with the encrusted spines.

FAMILY: ECHINOTHURIDAE: Echinothurids

Members of this family possess a hemispherical test up to 170mm in diameter which is much less rigid than in most sea-urchins. When removed from sea-water the test becomes even more flattened than under normal conditions. The single represented species is *Asthenosoma varium*, which is also the most venomous echinoid occurring in the Red Sea. Its short secondary spines are covered with a thin layer of skin, forming inflated venom sacs towards their ends. When a spine penetrates a victim the venom enters the wound and causes an extremely painful sting.

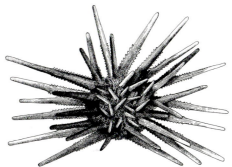

Prionocidaris baculosa

Phyllacanthus imperialis

Diadema setosum

Tripneustes gratilla

Tripneustes gratilla

Slate-pencil urchin: *Heterocentrotus mammillatus.*

FAMILY: DIADEMATIDAE: Diadematids

These urchins with their long, needle-sharp spines are a familiar sight in shallow sheltered areas, particularly around structures which create shade during daytime. They generally conceal (or partially conceal) themselves in holes or crevices during the day and emerge at night to browse on algal covered coral rubble or on other surfaces. There are four diadematids recorded from the Red Sea, i.e. *Chaetodiadema granulatum, Diadema setosum, Echinothrix calamaris* and *E. diadema.* Of these, *D. setosum* is perhaps the most abundant shallow water species. Its long black spines provide an effective defence against most predators, except for several persistent trigger and puffer fish which have been observed to attack them despite suffering numerous spine wounds! It is clear that their nocturnal manifestation is a behavioural adaptation which reduces the risk from predators, and it is interesting to note that in those sheltered areas where their main predators are relatively scarce, *D. setosum* may reach virtually plague numbers and they may then occur clustered into aggregations on open sea-grass meadows during daytime.

The spines of *Diadema* play a number of roles. In addition to their ability to point towards threatening objects and their painful effect when they penetrate and break off, thus warning many potential predators to keep clear, the spines also provide a sheltered niche for several small organisms.

A commensal anemone, *Coeloplana,* lives on the spines of *Diadema* and it has been shown that they move up the spines at night-time, thus attaining an improved position for capturing food particles transported by water currents.

Echinothrix diadema is superficially similar to *D. setosum* and has banded primary spines (as do juveniles of *D. setosum*) but adults have mainly dark spines. It has recently been recognised as a coral predator and can cause quite significant damage to corals (particularly *Porites, Montipora* and *Echinopora* species).

Adults of *Echinothrix calamaris* often retain dark and light banding on the spines and its spines have a much rougher feel when touched.

FAMILY: TEMNOPLEURIDAE: Temnopleurids

Temnopleurids are generally small urchins whose tests are usually characterized by distinct pits, troughs or pores at the angles where plates join each other. In some cases the plates themselves are more extensively sculptured. These features may be less apparent in *Microcyphus rousseaui,* which is distinguished by naked, dark purple areas around the joins of interambulacral plates on the upper (aboral) side of the cleaned test. It has reddish-brown spines with pale tips or in some cases the spines are banded purple and white. An additional feature is the presence of conspicuous spineless areas on the test.

Salmaciella dussumieri has primary spines which are banded green and white, and secondary spines which are bright vermilion. Other represented species include *Temnopleurus toreumaticus* and *T. scillae,* both of which appear to be confined to the Gulf of Suez.

Salmacis bicolor can reach a size of 100mm and has spines which are banded red and yellow or green.

FAMILY: TOXOPNEUSTIDAE: Toxopneustids

These urchins have smooth-surfaced tests without the sculpturing which occurs in the previous family. When seen from above the urchin test is circular or pentagonal. Around the edge of the peristome there are deep gill clefts. *Nudechinus gravieri* has a test which is not radially striped and primary spines with red or pink bands.

In *N. scotiopremnus* the spines are dark green around their bases, and paler or banded with green or brown distally. The test is green or spotted with green.

Tripneustes gratilla has an almost globular test in which only one in three or four ambulacral plates carry a primary tubercle. It is found among sea-grass beds in shallow water, where it frequently creates its own shade and camouflage by covering the upper surface of its test with sea-grass fronds, shell fragments and other pieces of debris which it picks up and holds in place with its tube-feet. There has been an apparent population explosion of this species in the Northern part of the Gulf of Aqaba where phosphate pollution has enhanced algal growth (consequently inhibiting corals), thus providing a rich food source for grazing sea-urchins.

FAMILY: PARASALENIIDAE: Parasalenids

Markedly elongate and generally small sea-urchins in which the test is flattened and periproct carries three to five plates. *Parasalenia poehli* has primary spines which are only about half as long as the test and are red and white banded.

FAMILY: ECHINOMETRIDAE: Echinometrids

Circular or ovate tests which lack pits or depressions at the angles between plates. Shallow gill clefts at the edges of peristome. Periproct with multiple plates.

Echinometra mathaei is probably the most abundant sea-urchin in the Red Sea and is found among corals and coralline rubble throughout the shallows. It has short, stout spines which are reddish-brown in colour with a small white ring encircling them basally. It is a fairly small urchin, generally about 5cms in diameter. *E. mathaei* actually excavates shallow depressions in dead coral by continually scraping the rock surface with its short spines.

Echinostrephus molaris takes this rock burrowing one stage further. This species has short black spines and lives on the reef top where they excavate small round holes which are inhabited on a permanent basis. In fact the urchins are often too big to crawl out of their burrows. Despite their imprisonment within the reef surface they are able to feed upon drifting algal fragments which are broken off by the strong wave-action which occurs across the reef-top. They capture the algae with long dorsal tube-feet and their protruding spines.

The slate-pencil urchin, *Heterocentrotus mammillatus* is another echinometrid urchin which is found in exposed situations along the reef crest and in shallow areas of fore-reef where wave-turbulence is strong enough to dislodge all but the most tenacious organisms. It is a bright, red-orange colour with broad, blunt-ended primary spines which are used to wedge the urchin into crevices where it hides during the day.

The related species *Heterocentrotus trigonarius* may be distinguished by the fact that its primary spines show a gradual decrease in size aborally instead of the abrupt shortening of those of *H. mammillatus*.

FAMILY: CLYPEASTERIDAE: Clypeasters

Clypeasterids are sand and gravel dwelling sea-urchins, which have flattened tests with a periproct close to the margin of the disc. Tube-feet on the aboral surface form distinct "petals". There are five genital pores in the test.

Represented species include: *Clypeaster amplificatus, C. humilis, C. rarispinus, C. reticulatus* and *C. fervens* (which is restricted to the Gulf of Aqaba). They are generally found just below the surface of the sand and are frequently collected in shrimp trawls.

FAMILY: FIBULARIIDAE: Fibularids

Fibularids are small sea-urchins, generally less than 2cms long with more or less ovate tests and poorly developed petals. The genus *Fibularia* has high, almost globular tests whereas those of *Echinocyamus* are flattened.

Represented species include: *Echinocyamus crispus* (which has a periproct with five or six regular radiating spineless plates), *E. elegans, Fibularia ovulum* and *F. volva,* which has a more flattened test than the globular form of *F. ovulum.*

FAMILY: LAGANIDAE: Laganids

The markedly flattened adult test is usually 3 to 5cms long, occasionally longer. In *Lagunum depressum* the periproct is closer to the posterior margin, whereas that in *L. joubini* is approximately midway

Echinometra mathaei

Typically round, excavated burrow of sea urchin *Echinostrephus molaris* on fringing reef platform in wave-breaking zone

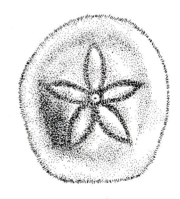

Lagunum depressum

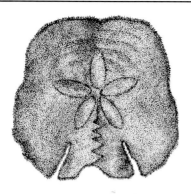

Echinodiscus auritus

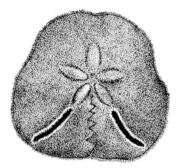

Echinodiscus bisperforatus

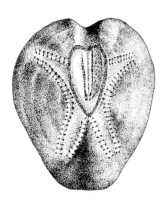

Lovenia elongata

Maretia planulata

between the posterior margin and the mouth. The petaloid area covers about half the test length in *L. joubini* and two thirds in *L. depressum*.

These sea-urchins superficially resemble *Clypeasters* but small spines on the upper surface have their tips expanded into crown shaped structures. They occur on sand or gravel and are usually found burrowing just below the surface.

FAMILY: SCUTELLIDAE: Scutellids
These have an extremely flattened test perforated by two or more lunules. They are often referred to as "sand-dollars".

In *Echinodiscus auritus* the lunules are open distally. This species is found sublittorally on exposed sandy beaches along both coastlines. *E. bisperforatus* has lunules which are at least as long as the petals. It has a similar habitat to *E. auritus*.

FAMILY: ECHINOLAMPADIDAE: Echinolampids
These have a peristome which is at, or only slightly anterior to, the centre of the oral surface. In *Echinolampas alexandri* the peristome is oval, while in that of *Palaeostoma mirabile* it is pentagonal.

FAMILY: LOVENIIDAE: Loveniids
Heart-shaped test with ambulacral areas petaloid and an inner fasciole present (see drawing). *Lovenia elongata* has a test whose posterior end has a deep funnel-like concavity. It lives in burrows in sandy areas.

FAMILY: SPATANGIDAE: Spatangids
These also have heart-shaped tests but unlike *Lovenia* there is no fasciole surrounding the petaloid area. *Maretia planulata* has a flattened test and is another sand burrowing species.

FAMILY: SCHIZASTERIDAE: Schizasterids
These are also heart-shaped sea-urchins in which the test displays fascioles around the petaloid area and in the latero-anal region, but not marginally. This is illustrated in the drawing of *Schizaster lacunosus*. Other represented species are *Diploporaster savignyi*, *Moira stygia* and *Paraster gibberulus*.

FAMILY: BRISSIDAE: Brissids
Brissids also have heart-shaped tests and a petaloid ambulacral area. A fasciole surrounds the petals and another one is present beneath the anus. Represented species include: *Brissopsis luzonica*, *Brissus latecarinatus*, *Metalia spatangus* and *M. sternalis*. They live in burrows in sand.

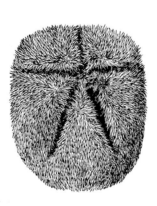

Brissus latecarinatus

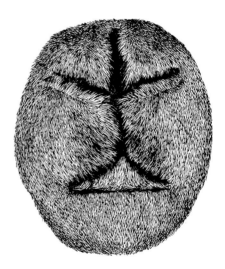

Metalia sternalis

Class: Holothuroidea: Sea Cucumbers

Sea cucumbers are significant members of the coral-reef ecosystem and it has been calculated that they are responsible for moving approximately 150 tons of sand per hectare per year (based upon an assumed population of 10,000 holothurians per hectare). They do this as part of their feeding activity in which they ingest sand along with their food and then eject undigestable material from the cloaca in the form of small faecal pellets. There is a wide variety of forms present, with around eighty species belonging to twenty-two genera having been recorded from the region. The most conspicuous species are members of the *Holothuriidae* and *Stichopodidae*. The main external features of a sea-cucumber are illustrated in figure 42. Identification of species often rests on the structure of their skeletal spicules.

Actinopyga mauritania

FAMILY: HOLOTHURIIDAE: Holothuriids

There are fifty species of Holothuriids recorded from the region. Three genera are represented, i.e. *Actinopyga*, *Bohadschia* and *Holothuria*. Most members of the family have thick body walls and many reach a length of 30cms or more. They have tube-feet and leaf-like oral tentacles. The gonads are present as a single tuft and spicules are diverse in form, lacking 'C' or 'S' shaped rods.

A complete listing of species is given in Table XXIV.

Actinopyga mauritiana: Surf Red Fish, grows up to 30cms in length and has tube-feet scattered all over the underside. It has a distinct bicolour pattern (deep reddish-brown above and cream or yellow below), and is armed with pointed spines on the upper surface. It occurs in *Halophila* sea-grass meadows and on sand close to coral reefs.

Bohadschia graeffei can be identified from the short, white-tipped papillae which occur on the darker areas of the dorsal surface. It is found on hard surfaces, frequently on dead corals encrusted by calcareous algae.

Holothuria atra (= Halodeima atra) is a common species (approximately 20cms long) which occurs in sandy areas, usually on sea-grass meadows where it can be observed on the surface of the sediment, occasionally camouflaged by mucus-adhered sand grains but more often than not quite conspicuous by its black sausage-like form contrasting against the light coloured sediment. It is used for "bêche-de-mer" but is not one of

Holothuria atra

Bohadschia graeffei

Table XXIV

RED SEA HOLOTHUROIDEA

HOLOTHURIIDAE

Actinopyga bannwarthi

A. mauritiana, A. miliaris

A. plebeja, A. serratidens

Bohadschia cousteaui

B. drachi, B. graeffei

B. marmorata, B. steinitzi

B. tenuissima

Labidodemas semperianum

Holothuria rigida, H. sucosa

H. atra, H. edulis, H. fungosa

H. massaspicula, H. insignis

H. pardalis, H. poli,

H. fuscocinerea, H. leucospilota

H. papillifera, H. pervicax

H. albiventer, H. brauni

H. martensi, H. ocellata

H. scabra, H. tortonesei

H. nobilis, H. difficilis

H. parva, H. cinerascens

H. flavomaculata

H. fuscoolivacea, H. hamata

H. klunzingeri, H. kurti

H. spinifera, H. squamifera

H. aphanes, H. arenicola

H. hilla, H. impatiens

H. remollescens, H. strigosa

H. aegyptiana, H. proceraspina

STICHOPODIDAE

Stichopus chloronatus

S. montotuberculatus

S. pseudhorrens

S. variegatus, S. sp.

CUCUMARIIDAE

Pentacta doliolum

P. pusilla

Pseudocnus echinatus

P. sp.

Stolus buccalis

Trachythone crucifera

T. dollfusi

PHYLLOPHORIIDAE

Ohshimella ehrenbergi

Phyllophorus calypsoi

Semperiella tenera

CHIRODOTIDAE

Chirodota stuhlmaanni

SYNAPTIDAE

Euapta godeffroyi

Labidoplax sp.

Leptosynapta chela

L. steinitzi

Opheodesoma grisea

O. kamaranensis

Patinapta crosslandi

P. dumasi

Polyplectana kefersteini

Protankyra autopista

P. pseudodigitata

Synapta maculata

Synaptula reciprocans

S. recta

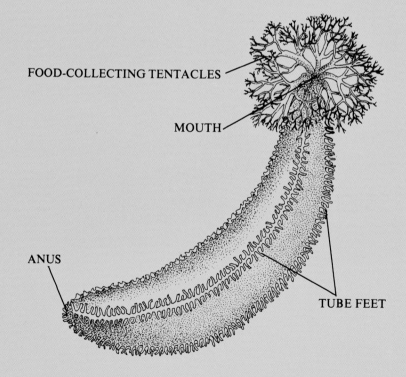

FOOD-COLLECTING TENTACLES

MOUTH

ANUS

TUBE FEET

Figure 42: *Generalized Sea Cucumber*

the most prized species for this purpose. It also burrows into the fine sand associated with sea-grasses. If provoked it may eject its guts but it lacks the well-developed cuverian tubules present in some closely related species.

Holothuria edulis (= Halodeima edulis) is a long, sausage-like sea-cucumber growing to 35cms or more. It has a thick soft skin and inconspicuous warts on a dark brown, grey or black dorsal surface, while the underside is tinged with deep pink or red. It lives on sand or among coral rubble. Despite its name it is little used as human food.

Holothuria impatiens is characterized by large warts on the dorsal surface and the skin has a gritty feel to it. It is found on coral sand and usually hides under rocks in daytime.

Holothuria leucospilota wedges itself under rocks and stretches out its front end over the sediment. If disturbed it can contract back into its shelter and readily ejects its sticky cuverian tubules.

Holothuria edulis

FAMILY: STICHOPODIDAE: Stichopodids

These large sea cucumbers usually lie on the surface of the sea-bed, and have long tubercles projecting from thick body-walls. Like holothuriids they possess tube-feet and have leaf-like oral tentacles. They differ however, in possessing two tufts of genital tubules and in the form of their spicules which include 'C' and 'S' shaped rods.

Stichopus chloronotus is a dark green, or even black, toxic sea cucumber which has warts (often red-tipped) on the dorsal surface. Its ventral tube-feet are arranged in three bands with the middle band widest. It is a relatively large sea cucumber (around 20cms long) and is found among coral rubble. Despite its toxicity it is occasionally eaten by sharks.

Stichopus variegatus

Stichopus variegatus is yellow-brown and frequently spotted on its dorsal surface, which bears low projections crowned by small papillae. It occurs among sand and shallow vegetation close to coral reefs. It is much larger than *S. chloronotus* and specimens as long as one metre have been found. It is favoured for use as bêche-de-mer.

FAMILY: CUCUMARIIDAE: Cucumariids

The cucumariids have ten bush-like tentacles. There are four genera and seven species represented (see Table XXIV). A species of *Pseudocnus* is often found in sand shaded by micro-atoll coral outcrops in shallow water.

FAMILY: PHYLLOPHORIDAE: Phyllophorids

Phyllophorid sea cucumbers have more than ten tentacles. Their spicules are tubular in form with a spine composed of three or four pillars. Represented species are listed in Table XXIV.

FAMILY: SYNAPTIDAE: Synaptids

These worm-like sea cucumbers lack tube-feet, but their skin remains "sticky" as a result of many anchor-like spicules being embedded in their skin. They can contract their bodies, which are soft and entirely flexible. They occur in coral pools and among shallow coral rubble as well as among sea-grass meadows. Represented species are listed in Table XXIV.

Synapta maculata

Thelenota ananas

Synapta maculata is a pale, fawn coloured synaptid which has an extended body length of up to two metres. When touched the skin feels sticky. It occurs among coral rubble in the shallows and is present on *Halophila* sea-grass beds. It feeds on organic matter sticking to the sand.

FAMILY: CHIRIDOTIDAE: Chiridotids

These small sea cucumbers lack tube-feet and have characteristic wheel-shaped spicules in their skin. They burrow in sand or gravel or among coral rubble. The single recorded species in the Red Sea is *Chiridota stuhlmanni*.

10. UROCHORDATES

PHYLUM: CHORDATA

SUBPHYLUM: UROCHORDATA: Tunicates

Class: Ascidiacea: Sea Squirts

Sea squirts may live as separate individuals, frequently in quite dense aggregations, or they may have a compound colonial structure formed by budding. The basic form familiar to many people, is that of a translucent, gelatinous, tubular sessile animal attached at its base and with two openings at its opposite end. They are major fouling organisms which grow in large clumps on ship's bottoms and on harbour or oil-rig pilings. There is a wide variety of forms within the class, ranging from the conspicuous tubular form described above to thin encrustations of some compound ascidians which may cover large surface areas. The two openings at the free-end of each individual are the buccal and cloacal siphons which permit water to pass through the animals. The sea-squirt body is enveloped in a test or tunic, which may be translucent or opaque and is sometimes quite thick, but with a texture which varies from gelatinous to tough and almost leathery or hard and similar to cartilage.

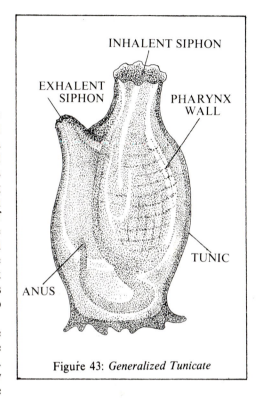

Figure 43: *Generalized Tunicate*

Water which is drawn in through the inhalent siphon passes to the anterior end of the gut and then through perforations (gill slits) into the exhalent chamber. The water current is created by cilia on the gill-slits. Sea squirts capture food particles born on this inhalent current by means of mucus. Bands of cilia transport food-laden mucus to the stomach.

The major families are discussed below, and a generalized illustration is shown in figure 43.

FAMILY: CLAVELINIDAE: Clavelinids

Colonial ascidians in which each zooid is relatively large, and often elongate. Budding occurs from the basal attachment stolons. The zooids are quite clearly differentiated into an abdomen and thorax, and the latter contains a pharynx (branchial sac) in which there are three to many rows of gill slits. In most cases the atrial siphons are separate but in some cases they open into a single common cloacal aperture. Representatives include: *Eudistoma praesslerioides* and *Stomozoa murrayi*.

Stomoza murrayi has club-shaped lobes (approx. 2cms in diameter) which are joined basally. The test is relatively hard and rigid and slightly purple in colour. The zooids occur away from the surface of colonies and careful dissection is necessary to reveal their structure.

FAMILY: POLYCLINIDAE: Polyclinids

Colonial ascidians which form massive colonies in which elongate zooids are deeply embedded within the tunic. There are more than seven rows of gill slits in the rather small branchial sac and each zooid may be subdivided into a thorax, abdomen and post abdomen.

They are found on the underside of coral boulders in the shallows and on exposed surfaces at greater depths. Represented species include: *Polyclinum saturnium; Amaroucium savignyi; A. erythraeum; Aplidium lobatum; Aplidium savignyi* and *Synoicum suesanum*.

Opposite, Top: Colonial ascidians; bottom: Tunicate on *Porites* sp., below: *Botrylloides* sp.

J. DAVID GEORGE

Botrylloides sp.

Didemnum moseleyi

Salp chain

Aplidium savignyi forms rounded lobes approx. 2cms in diameter which have a narrow base of attachment. The test is somewhat variable in form and may be soft and transparent so that zooids are completely visible.

FAMILY: DIDEMNIDAE: Didemnids
These colonial ascidians usually form flat, frequently brightly coloured encrusting sheets which spread over the substrate. The tunic contains spicules. Cloacal apertures are quite conspicuous since they are shared by several zooids within the colony, whereas inhalent siphons of the small individual zooids are inconspicuous. Represented species include: *Didemnum moseleyi; Didemnum candidum;* and *Diplosoma spongiforme.*

Didemnum moseleyi forms thin white colonies. The test carries many stellate spicules. Common cloacal apertures are conspicuous and open into a common cavity which divides the test into an upper, thin layer level with branchial siphons and a lower, thicker layer which encloses the abdomen. Branchial siphons are distinctly six-lobed.

FAMILY: ASCIDIIDAE: Ascidiids
Members of this family are comparatively large, solitary, ascidians in which the branchial siphon is eight lobed and the atrial one is six lobed. Representatives include: *Ascidia cannelata; Ascidia melanostoma; Ascidia nigra; Ascidia savignyi* and *Phallusia arabica.*

Ascidia melanostoma possesses a firm black test, approximately 9cms long by 5cms wide. Branchial aperture terminal and separate from atrial aperture. Can be easily confused with *Ascidia nigra* which can be separated by virtue of the fact that, unlike *A. melanostoma* it has separate openings of the dorsal tubercle into the branchial sac. It occurs in relatively deep water (i.e. 65m).

FAMILY: STYELIDAE: Styelids
This large family includes both solitary and colonial species. They are characterized by bodies which are not divided into abdomen and thorax and which possess a branchial sac with less than five folds on each side. Their siphons are smooth-edged or with four lobes. They have a cryptic habitat occurring under boulders or in crevices and on underhanging rock surfaces. Representatives include: *Styela canopus; Cnemidocarpa margaritifera; C. hartmeyeri; C. hemprichi; Polycarpa mytiligera; P. ehrenbergi; P. polycarpa; P. anguinea; P. coccus; P. steindachneri; Botryllus nigrum* and *B. rosaceus.*

FAMILY: PYURIDAE: Pyurids
Frequently quite large ascidians which are solitary, with a tough somewhat leathery tunic which may bear spines or tubercles. The branchial and atrial siphons each have four lobes. Representatives include: *Halocynthia spinosa; Pyura pantex; P. gangelion; P. sansibarica; Herdmania momus; Microcosmus pupa;* and *M. exasperatus australis.*

FAMILY: MOLGULIDAE: Molgulids
These are usually spherical in form and live partially buried in the sediment, not attached to a rock surface. The tunic may have a somewhat furry texture which enhances the "grip" of the ascidian to its surroundings. Siphons are six-lobed and curved gill slits are usually present in the pharynx. A represented species is: *Ctenicella clione.*

11. GLOSSARY

Abaxial	Surface which is remote from axis.	**Colline**	Small, hills, applied to corals.
Adambulacral	Structures adjacent to ambulacral areas in echinoderms.	**Columella**	Central pillar in coral skeletons or gastropod shells.
Adradial	In coelenterates, the radius midway between perradius and interradius.	**Corallite**	Cup of a single polyp of coral.
Ahermatypic	Non-reef building.	**Corallum**	Skeleton of compound coral.
Ambulacra	Locomotor tube-feet in echinoderms.	**Costae**	Anything rib-like in shape (e.g. as a ridge on coral or shell).
Anastomose	Union of five thread-like structures or tubes to form a network.	**Cuverian tubule**	Glandular tubes extending from cloaca of holothurians.
Asexual	Not sexual.	**Dentate**	Toothed.
Avicularia	In Bryozoa, a modified zooecium with muscular movable attachments resembling a bird's beak.	**Ecomorph**	Variety of a species found in a particular ecological niche.
Axial	Axis or stem.	**Explanate**	Spread out in a flattened extension.
Benthic	Living on the sea-bed.	**Exumbrella**	Upper, convex surface of jellyfish.
Branchiae	Gills of aquatic animals.	**Fasciole**	Ciliated band on certain echinoids for sweeping water of surrounding parts.
Byssus	Tuft of strong filaments secreted by a gland of certain bivalve molluscs. Used for attachment of mollusc to substrate.	**Foliaceous**	Leaf-like.
Calyx	Cup-like skeleton of corals.	**Fossa**	Pit or trench-like depression.
Capitate	Enlarged or swollen at tip, gathered into a mass at apex.	**Gastropod**	Mollusc with the ventral muscular disc adapted for creeping.
Capitulum	Part of column above the disc in sea-anemones.	**Gorgonian**	Horny coral with skeleton formed from gorgonin
Chelate	Claw-like or pincer-like.	**Hectocotylous**	One of the arms of a male cephalopod which is specialised to effect transfer of sperms.
Chelipeds	Forceps of decapod crustaceans; claw bearing appendages.	**Hemichordate**	Animals which possess a rudimentary notochord.
Chitin	The material which forms the skeleton of arthropods.	**Hydranth**	Feeding polyp (or zooid) in a hydroid colony.
Cilia	Hair-like structure which beats to create water currents.	**Hydrotheca**	Cup-like structure into which coelenterate polyp may withdraw.
Cloaca	Posterior end of intestinal tract in some invertebrates.	**Interradius**	Radius of a radiate animal halfway between two perradii.
Cnidosacs	Kidney shaped swelling or battery of stinging cells (e.g. on dactylozooids of Siphonophora).	**Lamellar**	Formed by or possessing thin plates.
Coenenchyme	Common tissue which connects the polyps of compound corals.	**Lappets**	Paired lobes extending downwards from distal end of stomodaeum in jellyfish; lobes of sea-anemone gullet.

Littoral	The littoral zone lies between high and low-water marks.
Lunules	Crescent shaped structure or marking.
Mandibles	Paired mouth-parts of crustaceans.
Massive	In biological terms this means bulky rather than necessarily large.
Maxillae	Paired appendages behind the mandibles in Crustacea.
Maxilliped	Appendages arranged in pairs behind maxillae in crustaceans.
Meandrine	Bears meandering convolutions, especially with regard to the surface of some corals.
Medusae	Soft, gelatinous body of a jellyfish.
Megasclere	Large skeletal spicules forming supporting framework of sponges.
Microsclere	Small spicules found scattered in sponge tissues.
Nematocyst	Stinging cell of coelenterates.
Nephridia	Excretory organs in invertebrates.
Oligochaete	A division of Annelida which includes earthworms and lug-worms.
Operculum	A lid or covering flap, usually used to close-off entrance of tube or shell after animal has withdrawn.
Ostia	Small, mouth-like opening (e.g. from flagellate canal into paragastric cavity in sponges).
Pallium	Molluscan mantle. Pallial line is that left on the shell as a result of the position of the mantle.
Papillae	Glandular hair or conical structure.
Parapodia	Paired lateral locomotory structures on body segments of polychaetes and also lateral extensions of the foot in Nudibranchs.
Pedicel	Foot-stalk or stem of fixed invertebrates.
Peduncle	Stalk of crinoids and barnacles; link between thorax and abdomen of crustaceans.
Pereiopod	Locomotory thoracic limbs of Malacostraca Crustaceans.
Periostracum	Chitinous external layer of most mollusc shells.
Periproct	Surface immediately surrounding arms of echinoids.
Perisarc	Tough outer membrane of Hydrozoa.
Peristome	Region surrounding mouth.
Petaloid	Like a petal.

Pharynx	Gullet leading from mouth to stomach.
Pinnules	Reduced parapodium in some polychaetes; in Crinoidea one of the side branches, two rows of which fringe arms.
Planula	Ovoid, free-swimming larva of coelenterates.
Planktivorous	Feeding on plankton.
Polyp	Separate zooid of coelenterate individual or colony.
Proboscis	Trunk like process on head of annelids, nemerteans etc.
Ramose	Branching.
Reticulate	Like network.
Rhinophores	Process on aboral side of eye of certain molluscs . . . with olfactory function, also chemoreceptor on anterior tentacle of some gastropods.
Rhopalia	Marginal sense organ of Discomedusae.
Scleractinia	Order of stony corals.
Setae	Bristle-like structure in annelids.
Sinistral	Coiling in anticlockwise direction.
Spinose	Bearing many spines.
Statocyst	Vesicle in many invertebrates which discerns position of body in space.
Stolon	Cylindrical stem of some Bryozoa, from which individuals grow out at intervals.
Styliform	Bristle-shaped.
Test	Shell or outer hard covering.
Theca	Protective capsule.
Tunic	Body wall or test of tunicates.
Velar	Near a velum.
Velum	In jellyfish, the annular membrane projecting inwards from margin of bell.
Ventral	Lower or abdominal side.
Verruca	Wart-like projections (e.g. on base of Alcyonarian polyps), or blister-like evaginations on body wall of some sea anemones.
Zooecium	The thickened and hardened part of the cuticle of each polyp in a colony of Bryozoa, forming a cell or sheath in which it is lodged.
Zooid	Each of the distinct beings which make up a colonial animal, especially in coelenterate and bryozoan colonies.
Zooxanthellae	Yellow or brown symbiotic algae which live in tissues of coral and other organisms.

12. REFERENCES

SPONGES

Bryan, P.G. 1973 Growth rate, toxicity and distribution of the encrusting sponge *Terpios* sp. (Hadromerida: Suberitidae) in Guam, Mariana Islands. *Micronesica 9* (2) 237 - 242.

Burton, M. 1952 The Manihine Expedition to the Gulf of Aqaba — Sponges. *Bull. Bri. Mus. Nat. Hist. Zoo. Vol. 1. No. 8.* 163 - 174.

Fishelson, L. 1966 *Spirastrella inconstans* Dendy (Porifera) as an ecological niche in the Dahlak Archipelago (Eritrea). *Bull. Sea Fish. Res. Stn. Haifa 41.*

Fishelson L. 1971 Ecology and distribution of the benthic fauna in the shallow waters of the Red Sea. *Marine Biology 10* (2) 113 - 133.

George, D. and J. *Marine Life* published by Harrap, 1979.

Keller, C. 1889 Die Spongienfauna des rothen Meeres 1. *Zeits. fur wiss. Zool. Bd. 48* 311 - 404.

Keller, C. 1891 Die Spongienfauna des rothen Meeres. 2. *Zeits. fur wiss. Zool. Bd. 52* 294 - 369.

Levi, C. 1965 Spongiaires récoltés par l'Expedition Israelienne dans le sud de la Mer Rouge en 1962. *Bull. Sea Fish. Res. Stn. Haifa. 40,* 13.

Mergner, H. 1979 Quantitative ökologische Analyse eines Rifflagunenareals bei Aqaba (Golfe von Aqaba, Rotes Meer). *Helgoländer wiss, Meeresunters 32* 476 - 507.

Mergner, H. and M. Mastaller 1980 Ecology of a reef lagoon near Aqaba (Red Sea). *Proc. Symp. Coastal & Marine Env. Red Sea.* Khartoum University Press Vol. 1. 39 - 76.

Patton, W.K. 1976 Animal associates of living reef corals. *Biology and Geology of Coral Reefs III* (2) 1 - 36. Academic Press.

Row, R.W.H. 1909 Marine Biology of the Sudanese Red Sea. Report on the Sponges collected by Mr Cyril Crossland 1904 - 5. Part 1. Calcarea *Journal of the Linnean Society XXXI* (208) 182 - 214.

Row, R.W.H. 1911 Report on the sponges collected by Mr Cyril Crossland in 1904 - 5. Part II. Non Calcarea. *Ibid.* 287 - 400.

Thomas, P.A. 1972 Boring sponges of the reefs of Gulf of Mannar and Palk Bay. *Proc. Symp. Corals and Coral Reefs — Cochin* pp. 333 - 362.

NOTE:— Additional general references on sponges are listed in George and George 1979 which also provides a classification and a number of useful illustrations.

HYDROZOA

Boschma, M. 1968 The Milleporina and Stylasterina of the Israel South Red Sea Expedition *Bull. Sea Fish. Res. Sta. Haifa 49* 8 - 17.

Briggs, E.A. and V.E. Gardner 1931 Hydroida *Great Barrier Reef Expedition 1928 - 29. Scientific Reports IV* (6) 181 - 196.

Mergner, H. 1967 Über den Hydroidenbewuchs einiger Korallenriffe des Roten Meeres. *Z. Mörph. Okol. Tiere* 60, 35 - 104.

Mergner, H. and E. Wedler 1977 Über die Hydroidpolypenfauna des Roten Meeres und seiner Ausgange. *Meteor Forsch. - Ergebnisse 24, 1 - 32.*

Mergner, H. 1977 Hydroids as indicator species for ecological parameters in Caribbean and Red Sea Coral Reefs. *Proc. 3rd Int. Coral Reef Symp.* Miami 119 - 125.

Millard, N.A.H. and J. Bouillon 1974 A collection of hydroids from Moçambique, East Africa. *Ann. S. Afr. Mus. 65* (1), 1 - 40, 9 figs.

Thornely, L.R. 1908 Hydroida collected by Mr C. Crossland from October 1904 to May 1905. *Marine Biology of the Sudanese Red Sea. Journal of the Linnean Society 31* (204) 80 - 85.

Vervoort, W. 1967 The Hydroida and Chondrophora of the Israel South Red Sea Expedition, 1962. *Bull. Sea Fish. Res. Sta. Haifa 43:* 18 - 54.

ALCYONARIANS

Dean, L.M.I. 1929 Report on the Alcyonaria. Zoological results of the Cambridge Expedition to the Suez Canal, 1924. *Trans. zool. Soc. Lond.* 22, 707 - 712.

Ehrenberg, C.G. 1834 Die Corallenthiere des Rothen Meeres. 156pp. Kgl. Akad. Wiss., Berlin.

Gohar, H.A.F. 1940 Studies on the Xeniidae of the Red Sea. *Publ. mar. biol. Stat. Ghardaqa.* 2. 25 - 118 pls 1 - 7.

Hickson, S.J. 1940 The species of the genus *Acabaria* in the Red Sea. *Publs. mar. biol. Stn. Ghardaqa* 2, 3 - 22.

Kükenthal, W. 1925 Octocorallia. Handb. Zool. 1, 690 - 796.

Klunzinger C.B. 1877 Die Korallenthiere des rothen Meeres. 1. Die Alcyonarien und Malacodermen. Gutmann, Berlin, 98pp.

Schuhmacher, H. 1973 Morphologische und ökologische Anpassungen von *Acabaria*-Arten (Octocorallia) im Roten Meer an verschiedene Formen der Wasserbewegung. *Helgoländer wiss. Meeresunters 25,* 461 - 472.

Stiasny, G. 1940 Gorgonaria aus dem Roten Meere Sammlung Dr Cyril Crossland, Ghardaga, und der "Mabahith" Expedition 1934 - 5. *Publ. Mar. Biol. Sta. Ghardaga.* 2 121 - 192.

Thomson, J.A. and Dean, L.M.I. 1931 The Alcyonacea of the Siboga Expedition, with an addendum to the Gorgonacea *Siboga Exped., Monogr.* 13d: 1 - 227 pls 1 - 28.

Thomson, J.A. and McQueen J.M. 1907 Reports on the marine biology of the Sudanese Red Sea. VIII. The Alcyonarians. *J. Linn, Soc. Lond. (Zool.) 31:* 48 - 75 pls 5 - 8.

Verseveldt J. 1965 Report on the Octocorallia (Stolonifera and Alcyonacea) of the Israel South Red Sea Expedition 1962, with notes on other collections from the Red Sea. *Bull. Sea Fish. Res. Sta. Haifa 40* (14) 28 - 48 pls. 3.

SCYPHOZOA

Gohar, H.A.F. and A.M. Eisawy 1961 The biology of *Cassiopea andromeda* (from the Red Sea) (with a note on the species problem). *Publ. Mar. Biol. Sta. Al-Ghardaqa 11* 3 - 44.

Halim, Y. 1969 Plankton of the Red Sea *Oceanogr. Mar. Biol. Ann Rev. 7* 231 - 275.

Maaden, H. van der 1958 Notes on *Aurelia aurita* (L.) and *Cassiopea andromeda* Eschscholtz from the Gulf of Aqaba. *Bull Sea Fish. Res. Sta. Haifa 12* 5 - 10.

Stiasny, G. 1937 Scyphomedusae. *Sci. Rep. John Murray Expedition* 1933 - 34. *Brit. Mus. IV* (7) 203 - 242.

Stiasny, G. 1938 Die Scyphomedusen des Rothen Meeres. *Verh. K. Akad. Wet. Amst.* (2) *37* (2), 1 - 35.

CORALS

Benayahu, Y. and Y. Loya 1977 Space partitioning by stony corals, soft corals and benthic algae on the coral reefs of the northern Gulf of Eilat (Red Sea). *Helgoländer wiss Meeresunters 30* 362 - 382.

Bertram, G.C.L. 1936 Some aspects of the breakdown of coral at Ghardaqa, Red Sea. *Proc. Zool. Soc. Lond. 1936* 1011 - 1026.

Crossland C. 1935 Coral faunas of the Red Sea and Tahiti *Proc. Zool. Soc. Lond. 1935* 499 - 504.

Crossland, C. 1938 The coral reefs of Ghardaqa, Red Sea *Proc. Zool. Soc. Lond. 1938* 513 - 523.

Ehrenberg, C.H. 1834 Beiträge zur physiologischen Kenntniss der Corallenthiere im allgemeinen und besonders des rothen Meeres nebst einem Versuche zur physiologischen Systematik derselben *Akad. Abh. preuss. Wiss* (Phys. math.) *1832* 225 - 380.

Fishelson, L. 1973 Ecological and biological phenomena influencing coral species composition on the reef tables at Eilat (Gulf of Aqaba, Red Sea) *Mar. Biol. 19* 183 - 196.

Forskål, P. 1775 Descriptiones Animalium edited by Carsten Niebuhr

Gardiner, J.S. 1909 The Madreporarian corals 1. The family Fungiidae. With an account of their geographical distribution. *Trans. Linn. Soc. Lond. 2nd ser. Zool. 12* 257 - 290.

Guilcher, A. and Berthois, L. 1955 Les recifs coralliens du nord du Banc Farsan, Mer Rouge. *Annls. Inst. Oceanogra. Monaco 30* 1 - 100.

Head, S.M. 1978 A cerioid species of *Blastomussa* (Cnidaria, Scleractinia) from the central Red Sea, with a revision of the genus. *Journ. Nat. Hist. 12* 633 - 639.

Head, S.M. 1980 The ecology of corals in the Sudanese Red Sea. *D. Phil. thesis Cambridge University.*

Klunzinger, C.B. 1877 Die Korallthiere des Rothen Meeres. Erste Theil: Die Alcyonarien und Malacodermen. Gutmannschen Verlag, Berlin.

Klunzinger, C.B. 1879a Die Korallthiere des Rothen Meeres. Zweiter Theil: Die Steinokorallen. Erster Abschnitt: Die Madreporaceen und Oculinaceen. Gutmannschen Verlag, Berlin 88pp.

Klunzinger, C.B. 1879b Die Korallthiere des Rothen Meeres. Dritter Theil: Die Steinkorallen. Zweiter Abschnitt: Die Astraeaceen und Fungiaceen. Gutmannschen Verlag, Berlin 100pp.

Loya, Y. 1972 Community structure and species diversity of hermatypic corals at Eilat, Red Sea. *Mar. Biol. 13* 100 - 123.

Loya, Y. 1975 Possible effects of water pollution on the community structure of Red Sea corals. *Mar. Biol. 29* 177 - 185.

Loya, Y. 1976 The Red Sea coral *Stylophora pistillata* is an r strategist. *Nature,* London *259* 478 - 480.

Loya, Y. 1976 Settlement, mortality and recruitment of a Red Sea scleractinian coral population. In *Coelenterate Ecology and Behaviour,* ed. G.O. Mackie. Plenum, New York pp89 - 99.

Loya, Y. 1976 Recolonization of Red Sea corals affected by natural catastrophies and man-made perturbations. *Ecology 57* 278 - 289.

Loya, Y. and L.B. Slobodkin 1971 The coral reefs of Eilat (Gulf of Eilat, Red Sea. *Symp. Zool. Soc. Lond. 28* 117 - 139.

Marenzeller, E. von. 1907 Expeditionen S.M. Schiff "Pola" in das Rote Meer. Nordliche und Sudliche Halfte, 1895/96 — 1897/98 Zoologische Ergebnisse 25. Tiefseekorallen. *Denkschr. Akad. Wiss. Wien. 80* 13 - 26.

Marenzeller, E. von. 1907 Ibid 27 - 97.

Mergner, H. 1971 Structure, ecology and zonation of Red Sea reefs (in comparison with South Indian and Jamaican reefs). *Symp. Zool. Soc. Lond. 28* 141 - 161.

Mergner, H. and H. Schuhmacher 1974 Morphologie, ökologie und zonierung von Korallenriffen bei Aqaba, (Golfe von Aqaba, Rotes Meer). *Helgoländer wiss. Meeresunters 26* 238 - 358.

Nesteroff, W. 1955 Les recifs coralliens du Banc Farsan Nord (Mer Rouge). Resultats scientifiques des campagnes de la Calypso 1. 7 - 54 *Annls. Inst. Oceanogr. Monaco 30.*

Rosen, B.R. 1971 The distribution of reef coral genera in the Indian Ocean *Symp. Zool. Soc. Lond. 28* 263 - 300.

Scheer, G. 1971 Coral reefs and coral genera in the Red Sea and Indian Ocean. *Symp. Zool. Soc. Lond. 28* 329 - 367.

Scheer, G. and C.S.G. Pillai 1983 Report on the stony corals from the Red Sea. *Zoologica* 1 - 198 41 plates.

Vaughan, T.W. and J.W. Wells 1943 Revision of the suborders, families and genera of Scleractinia, *Spec. Pap. Geol. Soc. Am. 44* 1 - 363.

Vine, P.J. and S.M. Head 1977 Growth of corals on Commander Cousteau's underwater garage at Shaab Rumi (Sudanese Red Sea) Saudi Arabian Natural History Society. Jeddah. Journal *1977* 6 - 18.

Wainright, S.A. 1965 Reef communities visited by the Israel South Red Sea Expedition 1962. *Bull. Sea Fish. Res. Sta. Haifa 38* 40 - 53.

Wood, E. 1983 *Corals of the World* TFH Publications 256pp.

ANNELIDA

Amoureux, L. et al 1978 Systematique et ecologie d'annelides polychetes de la presqu'il du Sinai. *Isr. Jour. Zool. 27* 57 - 163.

Banse, K. 1959 *Fabricia acuseta* n.sp.; *Fabriciola ghardaga* n.sp.; *Oriopsis armandi* (Claparède) aus dem Roten Meer (Sabellidae Polychaeta). Kiel. *Meeresforsch XV* (1) 113 - 116.

Ben-Eliahu, M.N. 1972 Polychaeta errantia of the Suez Canal, *Isr. Jour. Zool. 21* 189 - 237.

Ben-Eliahu, M.N. and J.Dafni 1979 A new reef building Serpulid genus and species from the Gulf of Elat and the Red Sea, with notes on other gregarious tubeworms from Israeli waters. *Isr. Journ. Zool. 28* 199 - 208.

Betz, K.H. and G. Otte 1980 Species distribution and biomass of the soft bottom faunal macrobenthos in a coral reef (Shaab Baraja, Central Red Sea, Sudan). *Proc. Symp. Coastal & Marine Environment of the Red Sea, Gulf of Aden and Tropical Western Indian Ocean.* Univ. of Khartoum. Vol 1. 13 - 38.

Day, J.H. 1965 Some Polychaeta from the Israel South Red Sea Expedition, 1962. *Isr. South Red Sea Exped. Reports. 38.*

Fauvel. P. 1957 Sur quelques Annélides Polychètes du Golfe d'Akaba. *Bull Sea Fish, Res. Sta. Haifa 13* 3 - 11.

Fishelson L. and F. Rullier 1969 Quelques Annélides Polychètes de la Mer Rouge. *Isr. J. Zool. 18* 49 - 117.

Gravier, M.C. 1908 Contribution a l'étude des annélides polychètes de la Mer Rouge. *Nouvelle Archives du Muséum 4me série, Tome 10,* 67 - 168.

Vine, P.J. 1972 Spirorbinae (Polychaeta: Serpulidae) from the Red Sea, including descriptions of a new genus and four new species. *Zool. J. Linn. Soc. 51* (2) 177 - 201.

Vine, P.J. and J.H. Bailey - Brock 1984 Taxonomy and ecology of coral-reef tube worms (Serpulidae, Spirorbidae) in the Sudanese Red Sea. *Zool. J. Linn. Soc. 80* 135 - 156.

ECHIURA and SIPUNCULA

Stephen, A.C. 1965 Echiura and Sipuncula from the Israel South Red Sea Expedition. *Israel S. Red Sea Exped. Rep. 40.*

Wesenberg-Lund, E. 1957 Sipunculoidae and Echiuroidae from the Red Sea. *Bull. Sea Fish Res. Sta. Haifa 14* 3.

CRUSTACEANS

Achituv, Y. 1972 The zonation of *Tetrachthamulus oblitteratus* Newman, and *Tetraclita squamosa rufotincta* Pilsbry in the Gulf of Elat, Red Sea. *J. Exp. Mar. Biol. Ecol. 8* 73 - 81.

Banner, D.M. and A.H. Banner 1981 Annotated checklist of the Alpheid shrimp of the Red Sea and Gulf of Aden. *Zoolog. Verhand.* Leiden. No. 190. 1 - 99.

Bruce, A.J. 1975 Coral reef shrimps and their colour patterns. *Endeavour XXXIV* (121) 23 - 27.

Bruce, A.J. 1976 Shrimps and prawns of coral reefs, with special reference to commensalism. in *Biology and Geology of Coral Reefs* edited by Jones O.A. and R. Endean. Vol. III (2) 37 - 94 Academic Press.

Bruce, N.L. and D.A. Jones 1978 The systematics of some Red Sea Isopoda (family Cirolanidae) with descriptions of two new species. *J. Zool. (Lond.) 185* 395 - 413.

Bruce, A.J. and A. Svoboda 1983 Observations upon some Pontoniine shrimps from Aqaba, Jordan. *Zoolog. Verhand.* No. 205 1 - 44.

Calman, W.T. 1927 Report on the Phyllocarida, Cumacea and Stomatopoda. Zoological Results of the Cambridge Expedition to the Suez Canal. 1924 XXVII *22* 399 - 401 1 fig.

Dart, J. et al 1973 Sudanese Red Sea Rock Lobster / Fisheries survey *Rep. of Cambridge Coral / Starfish Research Group.*

Griffin, D.J.G. and H.A. Tranter 1974 Spider crabs of the family Majidae (Crustacea: Decapoda: Brachyura) from the Red Sea. *Israel Journ. Zool. 23* 162 - 198. figs 1 - 4, 1 pl.

Guinot, D. 1962 Sur une collection de Crustacés Décapodes Brachyoures de mer Rouge et de Somali. *Boll. Mus. Civ. Sta. Nat. Venezia 15* 7 - 63 figs 1 - 37, pls. 1 - 4.

Guinot, D. 1967 La faune carcinologique (Crustacea, Brachyura) de l'Ocean Indien Occidental et de la Mer Rouge. Catalogue, Remarques Biogeographiques et Bibliographie. *Mem. Inst. Fond. Afr. Noires, 77* 235 - 352.

Guinot, D. 1967 Les espèces comestibles de crabes dans l'Océan Indien Occidental et la Mer Rouge. *Mem. Inst. Fond. Afr. Noire 77* 353 - 390 figs 1 - 8.

Halim, Y. 1969 Plankton of the Red Sea. *Oceanogr. Mar. Biol. Ann. Rev. 7* 231 - 275.

Heller, C. 1861 Beiträge zur Crustaceen Fauna des rothen Meeres. 1. *Sitz. - Ber. Akad. Wiss. Wien,* 43 (1) 297 - 374. pl. 1 - 4.

Holthuis, L.B. 1958 Hippidea and Brachyura (Dromiacea, Oxystomata and Grapsoidea). Crustacea Decapoda from the Northern Red Sea (Gulf of Aqaba and Sinai Peninsula) 11. *Bull. Sea Fish. Res. Sta. Haifa 17* 41 - 54 figs. 1 - 4.

Holthuis, L.B. 1958 Macrura. Crustacea Decapoda from the Northern Red Sea (Gulf of Aqaba and the Sinai Peninsula) 1. *Bull. Sea Fish. Res. Sta. Haifa 17* 1 - 40 figs. 1 - 15.

Holthuis, L.B. 1967 Some new species of Scyllaridae *Proc. Kon. Nederl. Akad. Wetensch. 70* (c) 305 - 308.

Holthuis, L.B. 1968 The Stomatopod Crustacea collected by the 1962 and 1965 Israel South Red Sea Expeditions. *Isr. Jour. Zool. 16* (1) 1 - 45. fig. 1 - 7.

Holthuis, L.B. 1968 The Palinuridae and Scyllaridae of the Red Sea *Zool. Meded. Leiden 42* (26) 281 - 301 pl. 1 - 2.

Holthuis, L.B. 1977 The Grapsidae, Gecarcinidae and Palicidae (Crustacea: Decapoda: Brachyura) of the Red Sea. *Isr. Jour. Zool. 26* 141 - 192.

Hughes, R.N. 1977 The biota of reef flats and limestone cliffs near Jeddah, Saudi Arabia. *J. Nat. Hist. 11* 77 - 96.

Ingle, R.W. 1963 Crustacea Stomatopoda from the Red Sea and gulf of Aden. *Bull. Sea Fish. Res. Sta. Haifa 33* 1 - 69 fig. 1 - 73.

Jones, D.A. 1974 The systematics and ecology of some sand beach isopods (Family Cirolanidae) from the coasts of Saudi Arabia. *Crustaceana 26* (2) 201 - 211.

Karplus, I. et al. 1974 The burrows of alpheid shrimp associated with gobiid fish in the northern Red Sea. *Marine Biology 24* 259 - 268.

Laurie, R.D. 1915 On the Brachyura. Reports on the marine biology of the Sudanese Red Sea XXI *Jour. Linn. Soc. (Zool.) 31* 407 - 475 fig. 1 - 5 pl. 42 - 45.

Lewinsohn, Ch. 1969 Die Anomuren des Roten Meeres (Crustacea Decapoda: Paguridea: Galatheidea, Hippidea). *Zool. Verhandl. Leiden 104* 1 - 213 figs. 1 - 37 pl. 1 - 2.

Lewinsohn, Ch. 1977 Die Ocypodidae des Roten Meeres (Crustacea Decapoda, Brachyura). *Zool. Verhand. Leiden 152* 45 - 84.

Lewinsohn, Ch. 1977 Die Dromiidae des Roten Meeres (Crustacea, Decapoda, Brachyura). *Zool. Verhand. 151* 1 - 41.

Magnus, D. 1960 Zur ökologie des Landeinsiedlers *Coenobita jousseaumei* Bouvier und der Krabbe *Ocypode aegyptiaca* Gerstaecker am Roten Meer. *Verh. deutsche Zool. Ges 1960* 316 - 329 figs. 1 - 12.

Magnus, D. 1967 Zur ökologie sediment bewohnender *Alpheus* Garnelen (Decapoda: Natantia) der Roten Meeres. *Helgoländer wiss Meeresunters 15,* 506 - 522.

Mergner, H. 1979 Quantitative ökoloische Analyse eines Rifflagunenareals bei Aqaba (Golfe von Aqaba, Rotes Meer) *Helgoländer wiss Meeresunters 32* 476 - 507.

Newman, W.A. 1967 A new genus of Chthamalidae (Cirripedia, Balanomorpha) from the Red Sea and Indian Ocean. *J. Zool. Lond. 153* 423 - 435.

Nobili, G. 1906 Faune carcinologique de la Mer Rouge. Décapodes et Stomatopodes. *Ann Sci. nat. Zool.* (9) 4 1 - 347.

Nobili, G. 1906 Diagnoses preliminaires de 34 especes et varietes nouvelle, et de 2 genres nouveau de Décapodes de la Mer Rouge. *Bull. Mus. Hist. Nat. Paris. 11* 393 - 411.

Ramadan, M.M. 1936 Report on a collection of Stomatopoda and Decapoda from Ghardaqa, Red Sea. *Bul. Fac. Sci. Egypt. Univ. 6* 1 - 43 pl. 1 - 2.

Safriel, C.U. and Y. Lipkin 1964 Note on the intertidal zonation of the rocky shores at Eilat. *Is. Jour Zool. 13* 187 - 190.

Schuhmacher, H. 1973 Das kommensalische Verhältnis zwischen *Periclimenes imperator* (Decapoda: Palaemonidae) und *Hexabranchus sanguineus* (Nudibranchia: Doridacea). *Marine Biology 22* 355 - 360.

Southward, A.J. 1967 On the ecology and cirral behaviour of a new barnacle from the Red Sea and Indian Ocean. *J. Zool, Lond. 153* 437 - 444.

Stebbing, T.R.R. 1909 On the Crustacea, Isopoda and Tanaidacea. *Reports on the Marine Biology of the Red Sea. Journ. Linn. Soc. Lond. (Zool.) 31* 215 - 230.

Zarenkov, N.A. 1971 On the species composition and ecology of the decapod Crustacea of the Red Sea (in Russian) *Benthos of the Red Sea shelf.* 155 - 203 figs. 63 - 88.

CHELICERATA

Stock, J.H. 1957 Contribution to the knowledge of the Red Sea Pycnogonida from the Gulf of Aqaba. *Bull. Sea Fish. Res. Sta. Haifa 13* (2).

Stock, J.H. 1958 The Pycnogonida of the Erythrean and the Mediterranean coast of Israel. *Bull. Sea Fish. Res. Sta. Haifa 16.*

Stock, J.H. 1964 Report on the Pycnogonida of the Israel South Red Sea Expedition. *Bull Sea Fish. Res. Sta. Haifa 35* (3).

MOLLUSCS

Note:— Mastaller, 1979 provides a recent review of Red Sea molluscan literature. The following is an abbreviated list of selected references.

Adam, W. 1942 Les Céphalopodes de la Mer Rouge. *Bull. Inst. Océan. 822* 1 - 20.

Adam, W. 1959 Les Céphalopodes de la Mer Rouge. Mission R. Ph. Dollfus en Egypte (1927 - 1929). *Res. sci.* 3e 28 125 - 193.

Adam, W. 1960 Cephalopoda from the Gulf of Aqaba. *Bull. Sea Fish. Res. Sta. Haifa 26* 1 - 26.

Elliot, C.N. 1908 Notes on a collection of Nudibranchs from the Red Sea. Reports on the marine biology of the Sudanese Red Sea. XI *Jour. Linn. Soc. Lond. (Zool.) 31* (204), 86 - 122.

Engel, H. and C.J. Van Eeken 1962 Red Sea Opisthobranchia from the coasts of Israel and Sinai *Bull. Sea Fish. Res. Sta. Haifa 30* 15 - 34.

Foin, T.C. 1972 The zoogeography of the Cypraeidae in the Red Sea Basin. *Isr. Journ. Malacol. 3* (1 - 4), 5 - 16.

Foin, T.C. & L.P. Ruebush 1969 Cypraeidae of the Red Sea at Massaua, Ethiopia with a zoogeographical analysis based on Schilder's regional lists. *The Veliger 12* (2) 201 - 206.

Franc, A. (1956) Resultats scientifiques des campagnes de la "Calypso" en Mer Rouge: IX Mollusques marines *Ann. Inst. Ocean. 32* 19 - 60.

Gohar, H.A.F. and G.N. Soliman 1963 On the biology of three coralliophilids boring in living corals. *Publ. Mar. Biol. Sta. Al. Ghardaqa 12* 99 - 126.

Goreau, T.F. et al 1969 On a new commensal mytilid (Mollusca: Bivalvia) opening into the coelenteron of *Fungia scutaria* (Coelenterata). *J. Zool. Lond. 158* 171 - 195.

Griffiths, R.J. 1953 Cypraeidae at Akaba, Jordan *J. Conch. 23* (10) 337.

Hoyle, W.E. 1907 On the Cephalopoda. Report on the marine biology of the Sudanese Red Sea. *Journ. Linn. Soc. London 31* 35 - 43.

Hughes, R.N. and A.H. Lewis 1974 On the spatial distribution, feeding and reproduction of the vermetid gastropod *Dendropoma maximum Journ. Zool. Lond.* 531 - 547.

Issel, A. 1869 Malacologia del Mar Rosso. Pisa 1 - 387.

Jickeli, C.F. 1874 Studien über die Conchylien des Rothen meeres. 1. Die Gattung Mitra. *Jahrb. Deutsch. Malakozool. Ges.,* 1: 17 - 54.

Kohn, A.J. 1965 Conus (Mollusca, Gastropoda) collected by the Israel South Red Sea Expedition 1962, with notes on collections from the Gulf of Aqaba and Sinai Peninsula. *Bull Sea Fish. Res. Sta. Haifa 38* 54 - 59.

Lamy, E 1905 to 1930 A series of papers published in *Bull. Mus. Nat. Hist.* Paris from 1905 to 1930 — see full list in Mastaller 1979.

Leloup, P. 1960 Amphineures du Golfe d'Aqaba et de la peninsula Sinai. *Bull. Sea Fish. Res. Sta. Haifa 29* (20) 25 - 55.

Marcus, E. and E. Marcus 1960 Opisthobranchia aus dem Roten Meer und von den Malediven. *Abh. Mathem. — Naturwiss. Klasse,* Wiesbaden. *12,* 871 - 934.

Mastaller, M. 1978 The Marine Molluscan Assemblages of Port Sudan, Red Sea. *Zool. Meded. 53* (13) 117 - 144.

Mastaller, M. 1979 Beiträge zur faunistik und ökologie der Mollusken und Echinodermen in den Korallenriffen bei Aqaba, Rotes Meer. *Doctoral dissertation. Ruhr Universitat Bochum.*

Mienis, H.K. 1970 A checklist of Terebridae from the northern part of the Red Sea, with notes on *Terebra areolata, T. consobrina* and *T. subulata. Journ. Isr. Malacol. Soc. 1* (2) 37 - 42.

Mienis, H.K. 1971a Strombidae (Mollusca, Gastropoda) collected by the Israel South Red Sea Expedition 1962 *Argamon 2* 87 - 94.

Mienis, H.K. 1971b Cypraeidae from the Sinai area of the Red Sea. *Argamon 2* 13 - 44.

Nasr, D.H. 1982 Observations on the mortality of the pearl oyster, *Pinctada margaritifera* in Donganab Bay, Red Sea. *Aquaculture 28* 271 - 281.

Schilder, F.A. 1965 The Cypraeidae of the Israel South Red Sea Expedition 1962. *Bull Sea Fish. Res. Sta. Haifa 40* 75 - 78.

Soliman, G.N. 1969 Ecological aspects of some coral-boring gastropods and bivalves of the north western Red Sea. *Amer. Zool. 9* 887 - 894.

Sharabati, D. 1985 *Red Sea Shells* 128pp. KPI.

Sturany, R. 1901 Lamellibranchiaten des Rothen Meeres. *Denkschr. math. — naturwiss. Cl.k . Akad. Wiss.* Wien 74 219 - 283.

White, K.M. 1951 On a collection of Molluscs, mainly Nudibranchs from the Red Sea. *Proc. Malac. Soc. London 28* (6) 241 - 253.

Yonge, C.M. 1967 Observations on *Pedum spondyloideum* (Chemnitz) Gmelin, a scallop associated with reef-building corals. *Proc. Malac. Soc. Lond. 37* 311 - 324.

BRYOZOA

Dumont, J.P.C. 1981 A report on the Cheilostome Bryozoa from the Sudanese Red Sea. *Journ. Nat. Hist. 15* 623 - 637.

Powell, N.A. 1967 Bryozoa (Polyzoa) from the South Red Sea. *Cah. Biol. Mar. 8* 161 - 183 pls. 1 - 3.

Powell, N.A. 1969 A checklist of Indo-Pacific Bryozoa in the Red Sea. *Isr. Jour. Zool. 18* 357 - 362.

Soule, J.D. and D.F. Soule 1974 The bryozoan - coral interface on coral and coral-reefs. *Proc. of the Second Int. Symp. on Coral Reefs 1* 335 - 340.

Thornely, L.R. 1912 The Marine Polyzoa of the Indian Ocean from H.M.S. Sealark *Trans. Linn. Soc. Lond. (Zool.) 2nd Series 15* 137 - 157 8 pls.

Waters, A.W. 1909 Reports on the marine biology of the Sudanese Red Sea XII. The Bryozoa Pt.1. Cheilostomata. *J. Linn. Soc. (Zool.) London 31* 123 - 181 pls 10 - 18.

Waters, A.W. 1910 Reports on the marine biology of the Sudanese Red Sea XV. The Bryozoa Pt 2. Cyclostomata, Ctenostomata and Endoprocta. *J. Linn. Soc. (Zool.) London 31* 231 - 256.

ECHINODERMS

Burfield, S.T. 1924 A new species and a new variety of ophiuroid with notes on a collection of the Ophiuroidea from the Sudanese Red Sea *Ann. Mag. Nat. Hist. 13* (9) 144 - 154.

Chadwick, H.C. 1907 Report on the Crinoidea of the Sudanese Red Sea. *Journ. Linn. Soc. London 31* 44 - 47.

Cherbonnier, G. 1955 Les Holothuries de la Mer Rouge. Résultats scientifiques des campagnes de la 'Calypso' V. *Ann. Inst. Ocean 30* 129 - 183.

Cherbonnier, G. 1963 Les Holothuries de la mer Rouge de l'Université Hébraïque de Jerusalem *Sea Fish. Res. Sta. Haifa Bull. 34* 5 - 10.

Cherbonnier, G. 1967 Deuxième contribution à l'études des Holothuries de la Mer Rouge collectés par des Israéliens *Bull. Sea Fish. Res. Sta. Haifa 43* 55 - 68.

Cherbonnier, G. 1979 Holothuries nouvelle ou peu connues de la Mer Rouge (Echinoderms) *Bull Mus. Hist. Nat. Paris* (4) 2 27 - 30.

Clark, A.H. 1937 Crinoidea Scientific Reports of the John Murray Expedition *4* (1) 87 - 108.

Clark, A.M. 1952 The 'Manihine' Expedition to the Gulf of Aqaba VII Echinodermata. *Bull. Brit. Mus. (Nat. Hist.) Zool Ser. 1* (8) 203 - 214.

Clark, A.M. and F.W. Rowe 1971 Monograph of shallow-water Indo-West Pacific Echinoderms. *Brit. Mus. (nat. Hist.) London* 234pp.

Clark, A.M. 1976 Echinoderms of Coral Reefs. In *Biology and Geology of Coral Reefs III* (2) Edited by O.A. Jones and R. Endean. *Academic Press.*

Clark, H.L. 1939 Ophiuroidea. Sci. Rep. John Murray Expedition. *6* 29 - 136.

Fishelson, L. 1971 Ecology and distribution of the benthic fauna in the shallow waters of the Red Sea. *Mar. Biol. 10* 113 - 133.

Fishelson, L. 1974 Ecology of the Northern Red Sea Crinoids and their Epi- and Endozooic fauna. *Mar. Biol. 26* 183 - 192.

Goreau, T.F. 1963 On the predation of coral by the spiny starfish in the Red Sea. *Bull. Sea Fish. Res. Sta. Haifa 35* 23 - 26.

James, D.B. and J.S. Pearse 1969 Echinoderms from the Gulf of Suez and the Northern Red Sea. *Journ. Mar. Biol. Assoc.* India. *11* 78 - 125.

Magnus, D.B.E. 1967 Ecological and ethological studies and experiments on the echinoderms of the Red Sea. *Studies Tropical Oceanography 5,* 635 - 664.

Mastaller, M. 1979 Beiträge zur faunistik and ökologie der Mollusken und Echinodermen in den Korallenriffen bei Aqaba, Rotes Meer. *Doctoral Dissertation, Ruhr Universität Bochum* 1 - 344.

Moore, R.J. 1985 A study of an outbreak of the Crown of Thorns starfish: *Acanthaster planci. Rep. Queen Mary College 1984 Red Sea Expedition to Dunganab Bay, Sudan.*

Mortensen, T. 1926 Zoological results of the Cambridge Expedition to the Suez Canal 1924 VI Report on the Echinoderms. *Trans. Zool. Soc. London 22* (1) 117 - 131.

Mortensen, T. 1939 Report on the Echinoidea of the Murray Expedition 1. Regular Echinoidea. *Sci. Rep. John Murray Exp. 6* 1 - 28.

Mortensen, T. 1948 Report on the Echinoidea of the Murray Expedition II. Irregular Echinoidea. *Sci. Rep. John Murray Exp. 9* 1 - 15.

Ormond, R.F.G. and A.C. Campbell 1971 Observations on *Acanthaster planci* and other coral reef echinoderms in the Sudanese Red Sea. *Symp. Zool. Soc. Lond. 28* 433 - 454.

Pearse, J.S. 1969 Reproductive periodicities in Indo-Pacific Invertebrates in the Gulf of Suez I. The echinoids *Prionocidaris baculosa* (Lamarck) and *Lovenia elongata* (Gray). *Bull Mar. Sci. 19* 323 - 350.

Pearse, J.S. 1969 (see above) II The echinoid *Echinometra mathaei* (de Blainville) *Bull. Mar. Sci. 19* 580 - 613.

Pearse, J.S. 1970 (see above) III The echinoid *Diadema setosum* (Leske). *Bull. Mar. Sci. 20* 697 - 720.

Price, A.R.G. 1982 Echinoderms of Saudi Arabia. Comparison between echinoderm faunas of Arabian Gulf, SE Arabia, Red Sea and Gulfs of Aqaba and Suez. *Fauna of Saudi Arabia 4* 3 - 21.

Roads, C.H. and R.F.G. Ormond 1971 New studies on the Crown of Thorns starfish *(Acanthaster planci)* from investigations in the Red Sea. *Camb. Coral Starfish Res. Group — UK 1971* 124pp.

Rutman, J. and L. Fishelson 1969 Food composition and feeding behaviour of shallow-water crinoids at Eilat (Red Sea). *Mar. Biol. 3* (1) 46 - 57.

Tortonese, E. 1936 Echinodermi del Mer Rosso. *Ann. Mus. Stor. Nat. Genova 59* 202 - 245.

Tortonese, E. 1959 Echinoderms from the Red Sea *Bull. Sea. Fish. Res. Sta. Haifa 29* (19), 17 - 22.

Tortonese, E. 1977 Report on Echinoderms from the Gulf of Aqaba (Red Sea) *Monitore Zoologico Italiano 12* (9) 273 - 290.

Tortonese, E. 1979 Echinoderms collected along the eastern shore of the Red Sea (Saudi Arabia) *Atti. Soc. ital. Sci. nat. 120* 314 - 319.

UROCHORDATA: ASCIDIACEA

Godeaux, J. 1960 Tunicies Pelagiques de Gulfe d'Eylath. *Bull. Sea Fish. Res. Sta. Haifa 29* 9 - 16.

Godeaux, J. 1963 Tunicies Pelagiques de Gulfe d'Eylath. *Bull. Sea Fish. Res. Sta. Haifa 34.*

Kott, P. 1957 The sessile tunicata, *Sci. Rep. John Murray Expedition 1933 - 34* Vol. X (4) 129 - 149.

Peres, J.M. 1960 Sur une collection d'Ascidies de la Cote Israelienne de la Mer Rouge et de la Peninsule du Sinai. *Bull. Sea Fish. Res. Sta. Haifa 30* 24 39 - 47.

Prudhoe, S. 1952 The Manihine Expedition to the Gulf of Aqaba. Turbellaria. *Bull. Britt. Mus. (Nat. Hist.)* Vol. 1. (8) 175 - 179.

13. ACKNOWLEDGEMENTS

This book could not have been written were it not for the research of many scientists whose published works on Red Sea invertebrates have provided the raw facts from which I have compiled a general account of the subject. I have made every effort to acknowledge these marine scientists, both in the main text and in the list of references. I take this opportunity to once again thank them for their own invaluable efforts in developing our knowledge of the invertebrate fauna of the Red Sea. In particular I wish to thank Professor Hans Mergner for his work on shallow-water ecology and Red Sea hydroids; Dr Georg Scheer and Dr C.S.G. Pillai for their invaluable review of Red Sea corals; Dr Stephen Head for his own lucid investigations into ecology of Red Sea corals; Dr Michael Mastaller for his ecological studies and particularly his research into molluscs and echinoderms of the region; Dr Ailsa Clark for her monograph on Red Sea echinoderms and Dr Andrew Price for a recent review on the zoogeography of the region's echinoderm fauna; Dr David Jones for his research on Red Sea isopods and for providing advice during the preparation of this text; Dr L.B. Holthuis for his research on Red Sea crustaceans and for his helpful correspondence during my years of Red Sea research; Dr A.J. Bruce for his research on Red Sea shrimps; Professor Banner and his wife for their own invaluable review of alpheids from the Red Sea; Dr Roger Hughes for his ecological studies and Dr Julie Bailey-Brock for her co-authorship of a paper on Red Sea Serpulidae.

The preparation by David and Jennifer George of "Marine Life" (published by Harrap) should not go without mention. Although its scope is not restricted to the Red Sea, many of the identified coral-reef species are from this area and their book has assisted many people to classify and identify local marine-life. In addition David George has kindly provided a collection of his underwater slides for use in this book and has freely given his advice whenever I have requested it. Another book which has been of considerable assistance to Red Sea divers, including myself, is that by Rupert Ormond and Gunnar Bemert (Red Sea Coral Reefs; published by Routledge and Keegan). Dr Ormond was also responsible for my own introduction to the Red Sea, as a member of the Cambridge Coral Starfish Research Group in 1970. I am grateful to him and to the group's originator Dr C.H. Roads for the opportunities which they provided for me to immerse myself in the Red Sea. Before leaving the topic of authors and their books I also wish to acknowledge the work of Doreen Sharabati (Red Sea Shells; KPI) and Dr Elisabeth Wood (Corals of the World; TFH). The sea-shell book by Doreen Sharabati is beautifully illustrated and well researched and belongs on every Red Sea lover's bookshelf. Elisabeth Wood's "Corals of the World" has, as its title suggests, a global coverage and the book includes underwater photographs together with skeletal illustrations of Indo-Pacific corals, many of which are present in the Red Sea. In addition she has kindly reviewed the slides of Red Sea corals which are used in the current publication and has greatly assisted in the task of identifying the illustrated coral species.

In the preparation of the chapter on Red Sea molluscs I was aided by Dr Horst Moosleitner who prepared an early draft and provided the list of recorded species. Dr Moosleitner also generously contributed a number of photographs.

There are very few marine biologists who have a good knowledge of field identification of Red Sea sponges. The author was particularly pleased to receive a contribution from Frank Nobbe who carried out his Ph.D research on Red Sea sponges, based mainly at Aqaba Marine Laboratory. He also provided a number of identified photographs which have greatly improved the scientific value of this section.

Among the many scientific and diving colleagues with whom I have enjoyed exploring the Red Sea, I should like to mention in particular Douglas Allen; Dr Stephen Head; Dr Dirar Hassan Nasr; Richard Moore; Dr Rupert Ormond; Dr J.E. Randall; Hagen Schmid; Georg Jungbauer and Captain Abdel Halim.

My years of research on both coastlines of the Red Sea would not have been possible without the financial support and research funding of a number of organisations including the University of Wales (Swansea University College); the University of Khartoum; Ministry of Overseas Development (UK); British Sea Fish Authority; Ministry of Agriculture and Water (Saudi Arabia) and the Saudi Arabian Department of Fisheries. I take this opportunity to thank all these bodies for their support.

I am most grateful also to the photographers who have provided such high quality photographs for use in the book. Their individual credits are separately listed.

For their willing assistance in preparation of the index I have pleasure to thank my daughters Triona and Sinead who sorted reference cards and my youngest daughter Megan who did her best to rearrange them into her own unique alphabet.

Finally, I wish to thank Jane Stark for her excellent illustrations and design work and Rafi Minhas for typing the manuscript.

While every effort has been made to ensure that the text is scientifically accurate, I wish to apologise for any errors which may have occurred. I should be pleased to hear from readers who feel they can add new information on Red Sea invertebrates so that the next edition can be improved upon.

14. INDEX